Friction Stir Welding
Dissimilar Aluminum Alloys

T0203753

Friction Stir Welding

Dissimilar Aluminum Alloys

Noor Zaman Khan,
Arshad Noor Siddiquee,
and Zahid A. Khan

CRC Press
Taylor & Francis Group
Boca Raton London New York

CRC Press is an imprint of the
Taylor & Francis Group, an **informa** business

CRC Press
Taylor & Francis Group
6000 Broken Sound Parkway NW, Suite 300
Boca Raton, FL 33487-2742

First issued in paperback 2020

© 2017 by Taylor & Francis Group, LLC
CRC Press is an imprint of Taylor & Francis Group, an Informa business

No claim to original U.S. Government works

ISBN 13: 978-0-367-57324-9 (pbk)
ISBN 13: 978-1-138-19675-9 (hbk)

Visit the Taylor & Francis Web site at
http://www.taylorandfrancis.com

and the CRC Press Web site at
http://www.crcpress.com

Contents

Preface

Friction stir welding (FSW) is indeed a solid state welding process that has enabled joining materials that are otherwise difficult to be welded by other fabrication processes. It is relatively a young material-joining technique, which was invented in 1991 and it has proudly celebrated its silver jubilee. The authors are associated and involved in this process since last several years and have been conducting extensive experimental and analytical investigations in this area. Many industries including automotive, shipbuilding, rail, aerospace, etc. are adopting FSW commercially, which has led to an ever-increasing involvement of FSW researchers and engineers in successful implementation of this novel fabrication process. For research or experimental work in any field, background knowledge of that field is essential for achieving success and therefore, there is a need for a basic understanding of the FSW process with some experimental examples that serve as the starting platform for senior students, scholars, and investigators and it has primarily motivated us to shape our FSW experimental works and their findings in the form of this treatise. FSW area is very wide and it is difficult to cover its various aspects in a single text. Therefore, only some important aspects of FSW have been covered in this treatise. This text intends to provide valuable information and data related to FSW of dissimilar aluminum alloys. Academicians, researchers, practicing welding engineers, metallurgists, and fabrication industries should benefit from the material presented in this work.

Chapter 1 presents an introduction to the subject, which includes demand of aluminum alloys in industries, joining of aluminum alloys, joining of dissimilar aluminum alloys, FSW of aluminum alloys, and FSW of dissimilar aluminum alloys. It also describes the importance and benefits of the current research work.

Chapter 2 describes the working principle of FSW process and historical background of the process with its advantages, disadvantages, and applications. It also discusses the tool design, FSW process parameters, machine for FSW, work fixture for holding workpiece during welding, and response measurement for defining weld quality.

Chapter 3 presents a study on friction stir welding of aluminum alloys, which explores problems related to the welding of aluminum alloys, and FSW of 2xxx, 5xxx, 6xxx, and 7xxx series aluminum alloys.

Chapter 4 provides the description on FSW of dissimilar aluminum alloys and it covers various issues related to dissimilar materials welding and major challenges in the friction stir welding of these dissimilar materials. It also focuses on dissimilar FSW of 5xxx–6xxx, 2xxx–7xxx, and 6xxx–7xxx series aluminum alloys.

Chapter 5 describes the methodology used for performing experimental study on FSW of dissimilar aluminum alloys (5083 and 6063), which includes experimental setup, machine used, and welding tool design. It explains the phenomenon of defect formation during FSW of dissimilar aluminum alloys and summarizes joining of dissimilar aluminum alloy using FSW.

Noor Zaman Khan
Arshad Noor Siddiquee
Zahid A. Khan

Authors

Noor Zaman Khan is a full time UGC sponsored BSR doctoral fellow in the Department of Mechanical Engineering at Jamia Millia Islamia (A Central University), New Delhi, India. His PhD in the area of friction stir welding/processing is in advanced stage of completion. He earned his master's degree in production and industrial engineering in 2013 from Jamia Millia Islamia and bachelor's degree in mechanical engineering in 2011 from Jawaharlal Nehru Technological University, Hyderabad. His major research interest includes materials structure–property correlation, welding engineering, nonconventional machining, machining, and optimization of design and process parameters using design of experiments. He has published more than 10 articles in reputed journals and conference proceedings so far.

Arshad Noor Siddiquee is professor in the Department of Mechanical Engineering at Jamia Millia Islamia (A Central University), New Delhi, India. He earned PhD and MTech from IIT Delhi. He has supervised several MTech dissertations and currently he is supervising 10 doctoral research scholars. His major research interest includes materials structure property correlation, materials processing, welding engineering, machining and optimization of design and process parameters. He has published more than 90 articles in reputed journals and conference proceedings. He has also coauthored four books related to engineering and one monograph as well.

Zahid A. Khan is professor in the Department of Mechanical Engineering at Jamia Millia Islamia (A Central University), New Delhi, India. He earned his PhD in 2001 from Jamia Millia Islamia, New Delhi, India. His major research interest includes optimization of design and manufacturing processes parameters, ANN and fuzzy modeling, environmental ergonomics, etc. He has supervised five doctoral research scholars and many MTech dissertations so far and currently he is supervising five doctoral theses. He has published more than 100 articles in reputed journals and conference proceedings so far. He has also coauthored four books related to engineering and two monographs as well.

CHAPTER 1

Introduction

1.1 INTRODUCTION

Welding is a joining process that fabricates various parts or components so as to produce products of complex shapes and geometry, which are otherwise too difficult to produce through other manufacturing processes. In order to produce efficient, compact complex products that can fulfill their functional and esthetic requirements, it is necessary to use a suitable fabrication process to assemble together several smaller components possessing exotic properties. Welding is a common option to join such components. Joining of dissimilar material often poses serious challenges to such an extent that joining is sometimes not possible at all. This problem is mainly because of difference in mechanical, physical, chemical, and metallurgical properties of the materials being joined. Difference in melting point, thermal expansion coefficient, thermal conductivity, etc. may cause failure at the weldments even during welding. Welding constitutes an essential manufacturing process that enables the production of a wide range of products being used in automotive, shipbuilding, aerospace, and several other industrial sectors. However, welding processes are extremely complex and multidimensional in terms of materials, process, and workmen skill, which make the fabrication of desired quality joint extremely difficult.

Joining of dissimilar materials with desirable overall quality is a challenging research field and welding of dissimilar materials has always been a matter of concern for engineers and scientists worldwide. There has been an ever-increasing demand for products possessing properties such as light weight, high strength, good corrosion resistance, etc. In order to fabricate a single structure, comprising several components often of different materials that exhibit various desirable properties, it is essential to join dissimilar materials together. Thus, welding of different grades of aluminum alloys having desirable mechanical and thermal properties owing to their high specific strength, thermal conductivity, and corrosion resistance are in great demand. Property–microstructure relationship in aluminum alloys is presented in Table 1.1. Several examples are found where aluminum alloys of different grades are joined together so as to provide desirable properties to the structure. For example, joining of 5xxx aluminum alloy (used for hull) with 6xxx aluminum alloy (used for secondary structural component) in a ship; similarly joining of 2xxx (a material for lower wing) and 7xxx series aluminum alloy (used to make upper wing) in aircraft (Figure 1.1), etc.

Economical and technical advantages of joining dissimilar materials have enabled its use in various industrial applications. Joining dissimilar materials by FSW has emerged as a new research topic. FSW has not only

TABLE 1.1   Property–Microstructure Relationship in Aluminum Alloys

Property	Microstructural Feature	Function of Feature(s)
Strength	Uniform dispersion of small, hard particles, fine grain size	Inhibit dislocation motion
Ductility and toughness	No large particles, clean grain boundaries, fine structure, no shearable particles	Encourage plasticity, inhibit void formation and growth, work harden
Fatigue crack initiation resistance	No shearable particles, fine grain size, no surface defects	Prevent strain localization and slip steps on surface, prevent stress concentration
Fatigue crack propagation resistance	Shearable particles, no anodic phases or hydrogen traps, large grain size	Encourage crack closure, branching, deflection, and slip reversibility
Pitting	No anodic phases	Prevent preferential dissolution of second-phase particles
Stress corrosion cracking, hydrogen embrittlement (HE)	No anodic phases, or interconnected hydrogen traps, hard particles	Prevent crack propagation due to anodic dissolution of HE, homogenize slip
Creep	Thermally stable particles on grain boundaries, large grain size	Inhibit grain boundary sliding

Source: Reprinted from *Progress in Aerospace Sciences*, 32, E.A. Starke, J.T. Staley, Application of modern aluminum alloys to aircraft, 131–172, 1996, with permission from Elsevier.

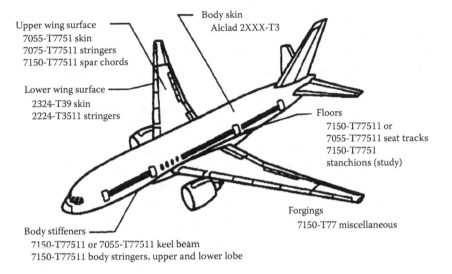

Upper wing surface
7055-T7751 skin
7075-T77511 stringers
7150-T77511 spar chords

Body skin
Alclad 2XXX-T3

Lower wing surface
2324-T39 skin
2224-T3511 stringers

Floors
7150-T77511 or
7055-T77511 seat tracks
7150-T7751
stanchions (study)

Forgings
7150-T77 miscellaneous

Body stiffeners
7150-T77511 or 7055-T77511 keel beam
7150-T77511 body stringers, upper and lower lobe

FIGURE 1.1 Application of different grades of aluminum alloys in Boeing 777. (Reprinted from *Progress in Aerospace Sciences*, 32, E.A. Starke, J.T. Staley. Application of modern aluminum alloys to aircraft, Copyright 1996, with permission from Elsevier.)

been found to produce near-defect-free joints with sound postwelding mechanical properties while joining various similar and dissimilar aluminum alloys but has also been able to effectively join a few previously difficult-to-weld aluminum alloys such as 2xxx and 7xxx series. However, to obtain acceptable quality welds important FSW process parameters need to be established for efficient joining of dissimilar aluminum alloys by preventing brittle intermetallic formation and imperfections in the joints to promote adequate flow of material and to mitigate deterioration in mechanical properties and surface morphology. Efficient and effective joining of dissimilar materials require adequate flow of material around the tool pin and proper mixing of material at stir zone (SZ) during welding for which the strategies pertaining to the joint design, tool design, and tool offset from the faying surface of base materials (BMs) need to be addressed as they play a critical role in the success of FSW of dissimilar alloys.

1.2 DEMAND OF ALUMINUM ALLOYS IN INDUSTRIES

Aluminum alloys possess various desirable properties such as good corrosion resistance, high strength-to-weight ratio, better fatigue strength that enable them to be used in different structural parts and other components for aerospace, marine, shipbuilding, and rail transport industries. The use of aluminum is expected to continue to increase worldwide, particularly in the transportation and manufacturing sectors. Aluminum alloys, being light in weight, have been the primary structural material for military and commercial aircraft for almost 80 years owing to their well-known mechanical behavior, strength-to-weight ratio, and mature manufacturing processes; and will remain so with the development of new-generation high-strength aluminum alloys. Use of light-weight material (aluminum alloys) in transportation sector reduces vehicle mass, which in turn minimizes fuel consumption and harmful emissions. Reduction in weight of the various modes of transportation reduces fuel consumption, which lessens frequent filling of fuel tanks. Use of light-weight material with high strength-to-weight ratio in making structures has a great impact on reduction in the cost that occur due to fuel consumption, frequent repair and maintenance, etc. Airframe manufacturers and material producers focus on the development of new aluminum alloys having good mechanical, metallurgical properties to meet customer requirements. Good mechanical properties and corrosion resistance of the materials may increase the life of the component and reduce repair costs.

Aluminum alloys are widely used by various industries in the fabrication of parts and components. More specifically 5xxx and 6xxx aluminum alloys have applications in shipbuilding, automobile, and aerospace, whereas 2xxx and 7xxx aluminum alloys have wide applications in aircraft components such as wings, tanks, fuselage, stringers, etc. as shown in Figure 1.1. Application of different aluminum alloys is listed in Table 1.2.

Reducing the weight of vehicles without compromising on the safety passengers are the two major challenges faced by automobile industries. Vehicle weight affects its performance, which is generally measured in terms of acceleration, top speed, and fuel consumption. Aluminum alloy is a light material with a high specific strength owing to which its use in the manufacturing of cars has tremendously increased. The use of aluminum alloy in space frame reduces the body weight of Audi A8 by 40% (Figure 1.2) (Miller et al., 2000).

Currently, all aluminum vehicles are also being produced on a commercial scale. Aluminum alloy sheets are widely used in inner and outer body panels of cars, which significantly reduce weight of vehicle. The sustained growth of industrial use of aluminum alloys depends to a great extent on the availability of a suitable joining process. Increasing use of aluminum in automobiles often requires dissimilar joining of steel with aluminum

TABLE 1.2   Specific Uses of Various Aluminum Alloys

Aluminum Alloys	Major Alloying Element	Typical Composition (wt.%)	Typical Properties and Application
1000 series	Unalloyed aluminum	>99 Al	Good electrical conductor, low strength: cooking foil, power transmission, utensils
2000 series	Copper	Al + 4–6 Cu + Mg	Strong heat-treatable alloy: aircraft external tanks, lower wings, fuselage
3000 series	Manganese	Al + Mn	Medium strength, excellent corrosion resistance, ductile: beverage cans, roofing, cooking pans, automotive radiators
5000 series	Magnesium	Al + 3 Mg	Strong work hardening alloy: pressure vessel, ship hulls, inner automotive body panel, boilers, storage tanks
6000 series	Magnesium + silicon	Al + Mg + Si	Moderate strength heat-treatable alloy: pipelines, bridges, external automotive body panel, structural members
7000 series	Zinc	Al + 6 Zn + 2 Mg + 1.5 Cu	Strong heat-treatable alloy: aircraft upper wings, fuselage
Al–Li alloys	Lithium	Al + 3 Li	Good strength to weight and low density: aircraft spar and skins

alloys, and employment of efficient joining techniques becomes highly crucial as these BMs have large differences in their physical, thermal, and chemical properties (Barnes and Pashby, 2000). A typical combination of strain hardenable Al–Mg (5xxx) alloys and the medium strength age hardenable Al–Mg–Si (6xxx) alloys is extensively used in automotive industry by car manufacturers. The 6xxx series alloys (e.g., AA6061) are exclusively used in external body panels and the 5xxx series alloys (AA5052) are used in inner body panels. But the biggest challenge with aluminum alloys is the problems associated with solidification during welding by conventional methods. Efficient welding process is required to weld the aluminum alloys so as to meet their heavy demand raised by user industries.

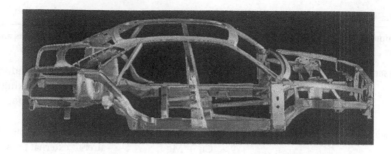

FIGURE 1.2   Aluminum space frame of Audi A8. (Reprinted from *Materials Science and Engineering A*, 280, W.S. Miller et al. Recent development in aluminum alloys for the automotive industry, 37–49, Copyright 2003, with permission from Elsevier.)

1.3 JOINING OF ALUMINUM ALLOYS

Despite several desirable mechanical properties possessed by aluminum alloys, they have not been able to completely replace other materials required by various industries. The major constraint that restricts the use of aluminum alloys is attributed to their joining process. Thus, novel joining techniques are required to efficiently weld them in order to fulfill the demand of user industries. Traditionally, mechanical fastening such as riveting, screwing, and occasionally arc welding had been used in fabrication of various parts for aircrafts and ships. However, mechanical fastening suffers from limitations such as it needs additional operations to maintain fit-up (i.e., creating holes and clamping, etc.), joints are prone to corrosion, and it is relatively difficult to make internal joints. Also, it acts as a crack initiation region in corrosive environment, which significantly reduces the joint strength (Barnes and Pashby, 2000).

Welding of aluminum alloys by fusion welding processes is difficult as compared to steel. Aluminum welding requires high heat input because of its high thermal conductivity and proper shielding gas due to high affinity to oxygen. Generally, aluminum alloys have melting point in the range of 570°C–650°C. Temperature requirement is high for achieving high heat input, which causes increase in the area of heat-affected zone (HAZ) that significantly deteriorates the quality of the welded joint. Also during welding of age hardenable aluminum alloys (2xxx, 6xxx, 7xxx) high heat input results in precipitate dissolution, which in turn degrades the mechanical properties. During welding of strain hardenable aluminum alloys (5xxx), high heat input results in loss of cold work, which in turn leads to reduction in mechanical properties. Moreover, relatively higher temperature during welding causes various defects such as porosity, solidification cracking, and weld distortion. Some fusion welding techniques such as high energy laser beam welding (LBW), high energy electron beam welding (EBW), and resistance spot welding (RSW) can be used for welding of aluminum alloys. However, several technical and economical limitations such as electrode wear in RSW, join line mismatch in LBW, high capital investment, etc. are associated with these techniques leading to the use of some other welding processes in place of these techniques (Barnes and Pashby, 2000).

1.4 JOINING OF DISSIMILAR ALUMINUM ALLOYS

Complex products are manufactured by assembling together different components often made of materials that differ in properties. Likewise, dissimilar welding is essential for making a product that comprises different aluminum alloys. Welding of dissimilar aluminum alloys has great potential in replacing riveted joints in aircraft and automotive structural parts. Over the past few years, aerospace, marine, rail, and automotive industry has developed an interest in joining dissimilar aluminum alloys. Successful welding of dissimilar materials is a challenging task due to the differences in chemical, physical, and metallurgical properties of the BMs. In fusion welding, filler metal is required and its composition depends on the nature of the BMs being welded and thus, it is difficult to choose filler material for fusion welding of dissimilar aluminum alloy. In addition, if one attempts to join dissimilar aluminum alloys by conventional fusion welding process, then there is a possibility of formation of several welding defects such as porosity, voids, hot cracking, distortion, etc.

Fusion welding requires melting of BM for obtaining permanent joint. Melting temperature and thermal conductivity of different aluminum alloys are different due to the presence of major alloying elements in them. Melting and solidification of welded material resulted in various problems associated with solidification microstructure that led to degradation of joint properties. Fusion welding of dissimilar materials is extremely difficult. However, difficulties can be overcome to some extent by using an interface layer of some compatible material so as to bridge the vast gap between physical, thermal, and mechanical properties of dissimilar materials. The interface material severely undermines the joint integrity and quality and hence use of fusion welding in a befitting application is not a feasible solution. Solid state welding overcomes the problem associated with conventional fusion welding processes. However, during solid state welding of dissimilar materials, considerable difference in the melting temperature of BMs affects their softening, which may lead to improper material mixing causing undesirable weld quality. Therefore, suitable solid state welding techniques (e.g., FSW) with proper strategy become very useful to effectively join dissimilar aluminum alloys.

1.5 FSW OF ALUMINUM ALLOYS

Aluminum alloys are widely used in several industrial sectors owing to their desirable properties such as strength, weight, corrosion resistance, etc. Individual parts of aluminum are produced separately and subsequently, they are joined together by means of effective joining techniques. Weldability of these alloys plays a vital role in the manufacturing of several products and assembly of structures. Joining of aluminum alloys is growing at a rapid rate as the demand for components made from them is increasing rapidly in automotive, marine, aerospace, and rail transport industries. To meet the ever-increasing demand of such products and structures, suitable joining techniques with maximum possible reliability and minimum costs are required for welding them. As stated earlier, conventional fusion welding processes involve melting of BMs, which gives rise to several problems associated with them leading to adverse solidification microstructure and joint quality. In addition, fusion welding processes deteriorate the properties possessed by BM due to several prewelding processes performed on them, particularly in the fusion welded zone and material in its neighborhood. The weldability of many aluminum alloys (2xxx and 7xxx series) is poor owing to unfavorable solidification microstructure, porosity, and hot cracking in the fusion zone and hence such aluminum alloys are not suitably joined by fusion welding processes. Melting and solidification of the fusion zone welds characterizes brittle interdendritic structure and it adversely affects the mechanical properties of the joint (Su et al., 2003). Unlike several fusion welding processes, FSW is a novel solid state joining process that does not involve melting and recasting of the materials being welded and thus overcomes the problems associated with fusion welding processes. In addition, it also encompasses several advantages; for example, it avoids/minimizes negative environmental impacts, conserves energy, it is safe for workers and consumers, and it is economically sound. FSW is the most significant development in materials joining in last 25 years.

FSW was initially invented and investigated for low melting point materials such as aluminum, magnesium, and copper alloys. Success of the

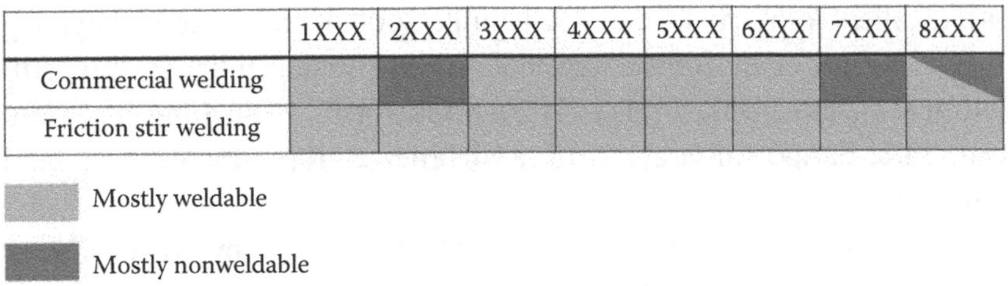

	1XXX	2XXX	3XXX	4XXX	5XXX	6XXX	7XXX	8XXX
Commercial welding								
Friction stir welding								

Mostly weldable

Mostly nonweldable

FIGURE 1.3   Weldability of aluminum alloys by conventional and FSW process.

FSW in effectively joining difficult-to-weld aluminum alloys (2xxx and 7xxx) by fusion welding gave a major Phillip for the phenomenal growth of this process. FSW can join any aluminum alloy effectively as depicted in Figure 1.3. Improved tensile and fatigue strength of the joints produced by FSW has increased the use of high-strength aluminum alloys in aerospace. FSW has replaced mechanical fastening (e.g., riveting) in aircraft structural parts (e.g., fuselage, wings) as sound joints of heat-treatable and nonheat-treatable aluminum alloys can be successfully obtained by FSW. For similar reasons, FSW finds increasing use in fabricating components of aluminum alloys in sectors such as automotive (Smith et al., 2001), aerospace (Nicholas and Thomas, 1998; Kallee et al., 2001), and railways (Kawasaki and Masai, 2004; Kallee et al., 2002) as well.

1.6 FSW OF DISSIMILAR ALUMINUM ALLOYS

Demand for simplicity in function, reduced weight, compact size, and esthetics has necessitated fabrication of discrete products from many components of different materials. Consequently, instances of industrial systems with an interface between different grades of aluminum alloys are common. Thus, dissimilar joining of aluminum alloys is required in several industrial applications in fabrication of structures, components, and transportation. Being a solid state welding, FSW is preferably suitable for the welding of aluminum and its alloys as it not only eliminates the problem of weld solidification cracking but also overcomes the problems associated with fusion welding processes. Hence, FSW has become a preferred fabrication process for dissimilar welding of aluminum alloys (DebRoy and Bhadeshia, 2010). The uniqueness of FSW of dissimilar materials has attracted tremendous research interest because of potential engineering and technical importance and problems associated with conventional fusion welding processes. In view of this, different dissimilar aluminum alloy combinations have been successfully friction stir welded with excellent joint efficiencies (Luijendijk, 2000; Khodir and Shibayanagi, 2008; Cavaliere et al., 2009; Priya et al., 2009; Sivashanmugam et al., 2010; Koilraj et al., 2012; Palanivel et al., 2012; Guo et al., 2014; Song et al., 2014; Aval, 2015; Ilangovan et al., 2015; Rodriguez et al., 2015; Venkateswarlu et al., 2015).

FSW is a solid state joining process that was invented by Wayne Thomas at The Welding Institute (TWI), Cambridge, in 1991. FSW utilizes heat generated by (i) friction between a nonconsumable rotating tool and workpiece and (ii) plastic deformation of the material being welded for joining materials together. FSW has successfully welded different combinations of aluminum alloy with good mechanical properties. Due to inherent potentials of FSW, it has become a process of choice for producing high-quality joints in both similar and dissimilar materials in just a short time span. In FSW, a special cylindrical tool comprising a pin at one of the shoulders is rotated and plunged into the faying surfaces of BMs with sufficient downward force as shown in Figure 1.4. After touching the tool shoulder on surface of BM, the tool traverses in welding direction along joint line. The rotating tool heats and softens materials under the shoulder and plastically deforms the softened material. Material moves from advancing side (AS) to the retreating side (RS) of tool and finally gets consolidated behind the tool pin by forging action of tool shoulder to complete the joint. Various zones are generated during traversing of tool in the welding direction as shown in Figure 1.5.

The demand raised by aerospace and other industries for components with reduced weight and associated production cost is increasing considerably. Efficient welding processes have proven to fulfill such demands by user industries (Rendigs, 1997). Use of components with reduced weight in the end products becomes important to lessen their operating expenses.

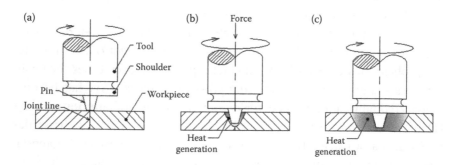

FIGURE 1.4 Plunging of FSW tool. (a) Rotating tool ready to plunge, (b) tool plunging into to the faying surfaces, and (c) plunging complete.

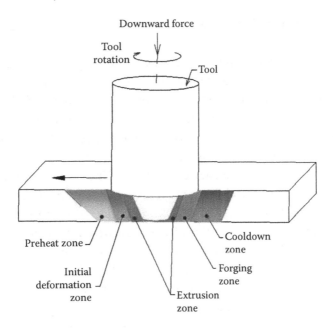

FIGURE 1.5   Various zones generated during movement of the FSW tool in welding direction.

The main aim of the manufacturers, especially aircraft manufacturers, is to minimize manufacturing cost without compromising on quality of products. This endeavor leads to significant saving in weight and cost, and increases the capacity per unit weight. The introduction and implementation of novel welding techniques is required to achieve such goals and FSW has proved to be an effective process, which has significantly reduced riveted/fastened joints and part count that were used in aircrafts earlier. Riveting requires drilling holes prior to joining that makes it a time-consuming process. Weight of rivet increases overall weight of the produced components. Use of FSW can replace riveting of aerospace components for saving weight and production time, which in turn may lead to cost savings. Reduction in weight helps in enhancing speed and also reducing fuel consumption. FSW is a novel cost-effective solid state welding technique that can join the materials effectively without adding any foreign material. It is employed in different industrial applications due to encouraging results observed in the performance of the joints. FSW, however, like other welding processes is not absolutely free from flaws, but fortunately, most of the flaws can be easily minimized by careful selection of process parameters. FSW is potentially used in joining large extruded profiles and panels that are widely used in shipbuilding and off-shore industry.

During FSW, heated and softened metals being joined undergo movement around the pin and under the shoulder by a stirring action. During stirring, the material experiences severe plastic deformation (SPD) to complete the joint. SPD causes the plasticized material to flow and the flowing material simultaneously undergoes dynamic recrystallization, which in turn results in the formation of ultrafine-equiaxed grains in the weld (Kazi and Murr, 2001; McNelley et al., 2008). Under correct combination of parameters the material being stirred consolidates behind the pin and produces defect-free joints with ultrafine-equiaxed grain structure and therefore, mechanical properties of the joint better than fusion welding are obtained. FSW has evolved as one of the best processes to successfully join dissimilar aluminum alloys and composites (Li et al., 1999a).

The importance of joining dissimilar materials is ever increasing in the pursuit of producing end products with reduced weight and improved performance. Several advantages offered by FSW, in comparison with other joining processes including fusion welding processes, has encouraged many industries such as automotive, aircraft, rail, and shipbuilding to adopt this novel process for joining dissimilar aluminum alloys. In order to exploit properties possessed by dissimilar aluminum alloys such as AA5xxx, AA6xxx, AA2xxx, and AA7xxx, there is a need to join their different combinations for which FSW is a better option than other joining processes.

Some of the major barriers of joining dissimilar aluminum alloys are as follows: different alloying elements, difference in metallurgical, thermal, and mechanical properties that hinder the material mixing and solidification imbalance leading to reduction in strength and integrity of a structure. Recent efforts on reducing such problems associated with welding of dissimilar aluminum alloys by FSW have led to the application and mass production of dissimilar aluminum alloys for industrial applications.

The progress made in dissimilar welding of light-weight materials such as aluminum and its alloys is the main enabler to the mass production of light transportation systems and consequently, significant reduction in fuel consumption. FSW has been adopted for joining aluminum alloys by the automotive industry for more than a decade. Recently, Honda Motor Corporation has employed dissimilar FSW of aluminum alloy with steel to produce a component to be used in the front structure of a Honda Accord (Figure 1.6). Honda is the pioneer in employing FSW for mass production

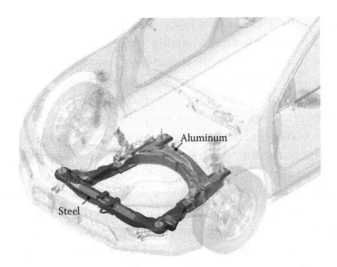

FIGURE 1.6 Honda front subframe joined by FSW. (From Honda develops new technology to weld together steel and aluminum and achieves world's first application to the frame of a mass-production vehicle, September 6, 2012. Available from: http://world.honda.com/.)

of vehicle subframe, which is a critical component of vehicle body frame. The front subframe of Honda vehicle is made of press formed steel halves and die cast aluminum and it supports the engine and some suspension components. To produce subframe, FSW was applied to weld the steel to the aluminum in a lap configuration at various locations. Honda claimed a reduction of 25% and 50% in the total body weight and electricity consumption, respectively, as compared to conventional all-steel subframe (Honda, 2012).

1.7 RATIONALE AND IMPORTANCE OF DISSIMILAR ALUMINUM WELDING USING FSW

FSW as a joining process, in general, and for joining dissimilar aluminum, in particular, has great potential in promising areas. Although in the current century there has been immense interest in FSW and significant progress has also been reported, yet it is far from being mature enough in terms of standardization, classification, and coding. In fact, the dissimilar FSW brings about great complexities due to variations in material properties, inequality of process forces on each material, intermetallic reactions, and material consolidation. All these issues bear a close relationship to main FSW process parameters, tool design, process strategies, etc., and consequently, make the task of fabricating good quality joints complex. This text has been created to thoroughly introduce the readers to the principle of FSW process and the phenomenon of joint formation during FSW of dissimilar aluminum alloys. Also, it focuses on various problems related to the welding of dissimilar aluminum alloys. It investigates the effect of various FSW process parameters on the joint quality as well as defect formation. The correlation between various important FSW process parameters and joint quality of dissimilar aluminum alloys is complex in nature. An experiment-based study of joining dissimilar aluminum alloys (5xxx and 6xxx) is presented in this monograph to explore effects of FSW process parameters on the joint quality. Influence of critical FSW parameters, that is, tool pin offset and plunge depth on microstructure and mechanical properties of the joint obtained by FSW of AA5083-H116 with AA6063-T6 (a combination typically used in ship building) has been demonstrated in the present monograph. This work focuses on the weldability and reliability of friction stir butt joints from the metallurgical point of view. Results of welding of dissimilar aluminum alloys obtained from mechanical tests and microstructure examination are presented for better understanding of the involved phenomenon. Various defects are also observed in welded joints and the effect of FSW parameters on defect formation is investigated and technical discussion regarding the same is also presented.

1.8 BENEFITS OF THIS TEXT

Good quality welds of dissimilar aluminum alloys are difficult to obtain using fusion welding processes due to problems associated with the solidification of microstructure of the joint. This problem degrades the postwelding mechanical properties of joint. Solid state welding processes can easily overcome such problems due to the absence of melting. Moreover, significant differences in properties of materials being welded such as thermal conductivity and melting point make the dissimilar material welding even more difficult.

FSW is very useful in joining dissimilar materials with sound postwelding properties due to lower processing temperature compared to fusion welding (Sato et al., 2004). Dissimilar aluminum welding finds tremendous industrial applications in fabrication of components that are subjected to complex loading conditions and corrosive environment. Thus, there is a need for an efficient welding process like FSW to join dissimilar materials to produce end products of light weight and high performance.

This monograph helps the reader in understanding principles of the process and also describes the methodology for performing research in the area of FSW. As a case study, dissimilar welding of aluminum alloys that have practical application in the area of shipbuilding is included.

This monograph is likely to provide the following benefits to its readers and researchers, both from academia and industries:

- It will provide a basic understanding of the FSW process, its parameters, and tool geometry.

- It will help in identification of the problems related to the welding of dissimilar aluminum alloys.

- It will make the readers understand the behavior of FSW of dissimilar aluminum alloys.

- It will facilitate understanding of the significance of process parameters and their effect on weld quality and defect formation.

- It will enlighten the readers with understanding of the microstructure evolution and characterization so as to develop their correlation with selected process parameters during FSW of dissimilar aluminum alloys.

It will equip the readers with various testing and measurement methods to evaluate mechanical behavior of the joints obtained by using different combinations of FSW parameters.

CHAPTER 2

Friction Stir Welding Process

2.1 INTRODUCTION TO SOLID STATE WELDING

Solid state welding can be performed at room temperature and also at elevated temperature without melting the materials being joined. In solid state welding, metallurgical bond is created below the melting point of material through plastic deformation without adding filler material (Messler, 1999). The energy supplied during solid state welding is through pressure and/or friction. The faying surfaces of materials are brought in intimate contact, required for direct atomic bonding, by expelling the surface contaminants through application of heat and/or pressure. Heat input in solid state welding is considerably lower than in fusion welding processes, which in turn causes less disruption of microstructure of parent materials being joined. Moreover, dissimilar materials can also be easily joined as no mixing of materials takes place in liquid form and therefore, sound-welded joints can be obtained. Various welding processes such as ultrasonic welding, cold welding, friction welding (FW), explosive welding, diffusion welding, and FSW fall under the category of solid state welding.

2.2 PRINCIPLE OF SOLID STATE WELDING

Solid state welding works on the principle of interatomic bonding obtained in solid state. Bonding force (interatomic force) between metallic atoms depends on their interatomic distance. By increasing interatomic distance to a few atomic spacing, the attractive interatomic force reduces to almost zero. Similarly, the force increases sharply and attains a very large value when the distance is reduced. If the faying surfaces of parent material to be welded are close enough to a value such that only grain boundary separates them, then parent material adheres with a very large force, resulting in a permanent joint. Solid state welding processes are characterized by the involvement of plastic deformation, which removes impure layer from the material's surface so as to bring pure atoms close enough to generate large attractive force to obtain permanent joint.

2.3 INTRODUCTION TO FSW TECHNIQUE

FSW process is a welding process in which the term *friction* refers to the utilization of frictional heat required for softening the BM and term *stir* signifies the movement of the material in the form of plastic deformation. Overall, FSW is a welding process that utilizes heat caused by friction between the tool and BM, and plastic deformation of BM caused by stirring of the tool. Generated heat softens the BM while plastic deformation mixes the BM that leads to the sound welded joint. Frictional heat was initially used in FW for joining materials. In conventional rotary FW process, one of the parts being joined is rotated while the other remains stationary. FW is limited to join cylindrical- (rotary friction) and rectangular-shaped components (for linear friction) of specific length; for example, small- and medium-sized round bars, tubes, and rectangular blocks. It is restricted to specific joint design and component geometry. Unlike FW, the FSW utilizes frictional heat for joining materials in different joint configurations.

FSW is a solid state welding process invented and patented by W. M. Thomas (Thomas et al., 1991) at "TWI" in Cambridge, UK. FSW technique was initially invented for welding of high-strength aluminum alloys (2xxx and 7xxx) (Dawes and Thomas, 1996), but its success made way for its phenomenal growth and it emerged as a major process for joining magnesium, copper and their alloys, ferrous, and other nonferrous alloys. It has been extended to dissimilar welding of the above-mentioned alloys and also to the welding of high melting point materials such as steel and titanium. Being a solid state welding process, FSW produces joints without melting of BM and therefore, problems associated with solidification are not encountered and joints made are free from porosity or blowholes leading to improved mechanical properties of the joints as compared to conventional welds. Various attempts have been made to implement FSW for joining dissimilar aluminum alloys (Guo et al., 2014; Aval, 2015;

Doley and Kore, 2016) and those materials having large difference in properties (physical, thermal, and mechanical) such as aluminum with steel (Liu et al., 2014), aluminum with copper (Akinlabi et al., 2014), and aluminum with titanium (Li et al., 2014). FSWed joints have largely replaced the use of riveted joints in aerospace because of their lower production costs and as a remedy for various problems associated with riveting. Therefore, FSW process has been identified as key technology for joining materials in fuselage and wing manufacturing by leading aircraft manufactures. Despite being a new process, products with welded joints made by FSW have already been launched into space in the form of seams in fuel tanks of a Boeing Delta II rocket in 1999.

FSW has now established itself to be a remarkable solid state welding technique to effectively join similar and dissimilar aluminum alloys. The process does not require any consumables (filler material, fluxes and shielding gas, etc.) for joining, produces no harmful emissions, safe to humans and is, therefore, considered to be an energy-efficient, environment-friendly, and clean material joining process, as detailed in Figure 2.1.

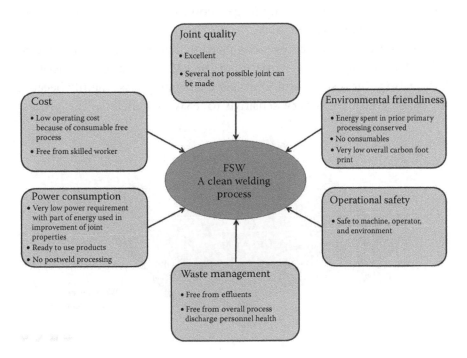

FIGURE 2.1   FSW, a clean welding process.

FSW can be used to obtain various types of joint configurations such as butt (Khan et al., 2015a), lap (Song et al., 2014), T-joint (Silva et al., 2014), fillet shapes, etc. Butt and lap joints are the most convenient joint configurations for FSW. A typical butt joint configuration is shown in Figure 2.2. Two rectangular plates or sheets are placed on the anvil of fixture in such a way that the faying surfaces of BMs touch each other. BMs are clamped by a robust clamping arrangement on work fixture to restrict their movement during plunging of the FSW tool and subsequent welding. Backing plate is used between BM and anvil to prevent sticking of the welded plates with the anvil of fixture. Large axial forces are applied by the tool during its plunging into BM and proper clamping of the BM is ensured so that the faying surfaces do not move apart. To complete weld, the rotating tool is plunged into joint line and traversed along welding direction keeping the tool shoulder in contact with the surface of material being welded. FSW involves step-by-step operations from initiation to completion of welded joints as shown in Figure 2.3. During welding, the material ahead of the traversing tool first starts preheating before the tool reaches the location and once the tool reaches there the material deforms plastically. Material in contact with the pin is extruded around it and forged by the tool shoulder behind it. Material already forged cools down after the tool moves ahead resulting in a welded joint.

FSW is a continuous autogenous process that involves a nonconsumable rotating tool, which is harder than the BM. Despite several merits of FSW, fabricating a successful joint is a challenging task, mainly because the quality of joint is very sensitive to FSW parameters. The process

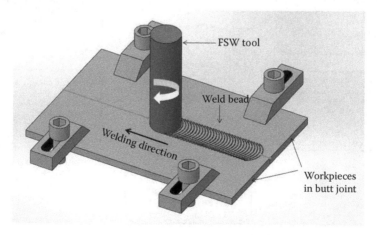

FIGURE 2.2   FSW of butt joint configuration.

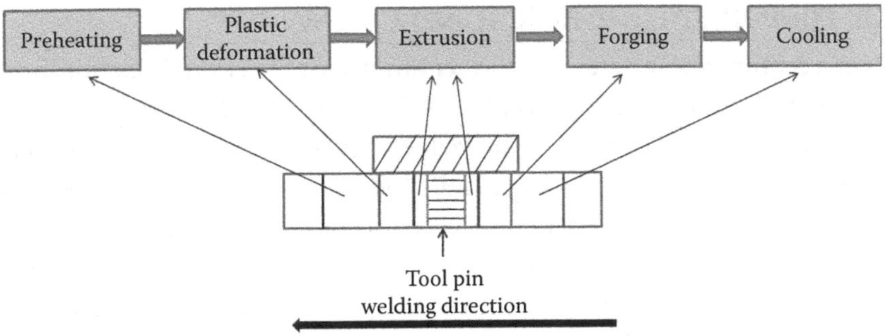

FIGURE 2.3 Processes involved in FSW.

requires careful selection of welding parameters, mainly tool rotational speed and traverse speed, tool plunge depth, tool tilt angle, tool offset, and tool geometry so that a defect-free joint is obtained (Leal and Loureiro, 2004; Khan et al., 2015b). FSW has undergone massive commercialization by many industrial sectors and has evolved as widely used solid state welding process due to superior joint properties relative to a conventional fusion weld.

It is important to understand the effects of FSW parameters on material properties, while determining the quality of welded joint. During FSW, the material undergoes SPD. The stresses and strains involved during SPD are complex to analyze and estimate due to the heterogeneous movement of material around the tool pin. FSW process affects the material not only thermally but also mechanically, leading to the creation of three microstructurally distinct regions of a friction stir weld: the SZ, the thermomechanically affected zone (TMAZ), and the HAZ. Each of these regions has unique microstructure and mechanical properties. Evolution of microstructure and mechanical properties in each of these regions of the joint produced by FSW are essential to understand and analyze.

Despite enormous merits, FSW is no exclusion to limitations. Capital investment in FSW infrastructure is one of its major disadvantages, but fortunately it can be significantly reduced by adapting heavy duty vertical milling machines to enable it to perform FSW. Based on the unique properties of the joints produced, commercialization of FSW began long before establishing a basic understanding of the process. Research on FSW is progressing at a very high pace in many industries especially in transportation industry. It is of great importance to develop deep insight into the effect of process and other parameters on the quality and cost effectiveness of the joint so that its potential can be exploited to the fullest.

2.4 HISTORICAL BACKGROUND OF FSW

Over a century ago, a patent on the very first use of frictional heat for form-ing and solid state welding was obtained by the United States (Bevington, 1891). Another development in the technology related to friction took place in the form of friction surfacing in a British patent in 1941 (Klopstock and Neelands, 1941). After another 50 years, a novel solid state welding was introduced that used frictional heat for welding operation. This invention is known as FSW, which was pioneered by "TWI" at Cambridge in 1991 (Thomas et al., 1991). This recent innovation has enhanced the use of tech-nology related to friction for producing welded joints in materials that are difficult to be welded such as high-strength heat-treatable aluminum alloys by conventional processes.

FSW is a revolutionary version of FW process, which can weld materi-als in different configurations such as lap, butt, T-joint, scarf, etc. with good postweld properties and little weld distortion. Moreover, FSW is a clean, economical, and simple welding process.

Initially, FSW mainly remained as a joining process for aerospace and shipbuilding applications. Now FSW has revolutionized the manufactur-ing sector and thereby commercialized in railways, automotive, and elec-tronics industries too. With optimum process parameters, FSW produces defect-free joints with good mechanical properties and little distortion in various similar and dissimilar materials that were previously not weld-able. As a solid state joining process, FSW has been a prominent process for welding dissimilar aluminum alloys. Various industries throughout the world (Japan, USA, Scandinavia, etc.) are using FSW as an important joining process, especially for welding of high-strength aluminum alloys.

2.5 FSW PROCESS

FSW process works on the heat produced by friction and plastic defor-mation affected by rotating tool. In FSW, BMs being welded are clamped together rigidly and a cylindrical nonconsumable tool with shoulder and pin is rotated and plunges into faying surfaces of the materials to be welded until the tool shoulder makes contact with the surface. Rubbing action of tool generates sufficient heat to soften BM and traversing tool mixes the material; thereby, completing the joint as shown in Figure 2.4. Tool rota-tion softens material under the shoulder and traversing tool deforms the material plastically, pushing it simultaneously from AS to RS and forges it behind the tool to complete joint. The tool shoulder prevents plasticized

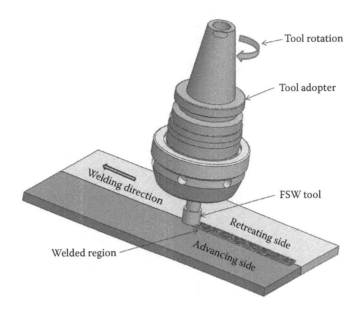

FIGURE 2.4   Schematic representation of FSW process.

material from being expelled out of the weld. Thus, the BMs are mechanically mixed while simultaneously undergoing SPD without melting. SPD due to stirring action of the tool probe/pin leads to dynamic recrystallization of the BM, which results in grain refinement and improvement of its properties.

During FSW, when the softened BM moves around the pin without reaching its melting point it avoids difficulties that arise from change in state (like melting, recasting, and volumetric changes), which are common in conventional fusion welding processes. Also, lower distortion and residual stresses generated due to lower welding temperature during FSW result in improvement in fatigue and fracture toughness that make thin material welding possible. FSW is a mechanized process with high equipment cost compared to conventional arc welding processes and less-skilled operator is required.

Before performing FSW, the material is first rigidly clamped on a welding fixture, as shown in Figure 2.5. Base plates are fixed in such a way that faying surfaces are close enough at the joint line so that the plates do not spread or lift during welding. Fixture design considerations play a vital role in the success of welding. This will be discussed later in detail under the heading "FSW fixture."

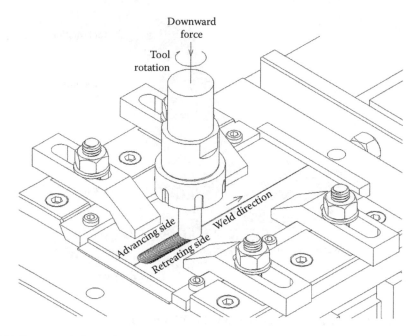

FIGURE 2.5   Robust clamping setup for FSW.

The process of FSW can be performed in a definite sequence of operations to complete the weld. These operations are described in the following four steps and the schematic representation is shown in Figure 2.6.

Step 1: Tool and workpiece clamping. The plates/sheets of required thickness are clamped together with supporting backing plate on the anvil of fixture supported by rigid clamps on the machine table/bed. The tool is clamped firmly on to the tool adopter, aligned along the joint line and rotated as shown in Figure 2.6a.

Step 2: Create friction/starting at edge. In this step, the weld is initiated and the tool pin is plunged vertically into the abutting surfaces of the BM along the joint with a specific force until the shoulder makes contact with the surface of the BM, as shown in Figure 2.6b. Stirring action of the tool softens the BM due to the frictional heat. Sufficient heat should be generated since a part of the generated heat is conducted to the anvil and the surrounding material.

Step 3: Joining/move tool when metal softens. After the welding tool is plunged into the BM, the tool traverses along the joint line in the

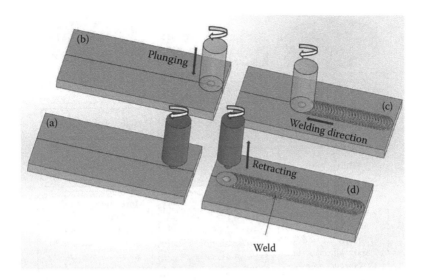

FIGURE 2.6   Four steps of FSW process.

welding direction with some dwell to allow the BM to reach sufficient temperature, as shown in Figure 2.6c. During traversing of the welding tool, heat produced by friction and plastic deformation maintains proper softening to allow sufficient material flow around the tool pin. Tool pin moves the plasticized material from AS to the RS and the tool shoulder consolidates the flowing material behind the pin to complete the joint.

Step 4: Plunging out of the tool. The process is completed when the rotating tool after joining the BM is retracted from the abutted surfaces leaving a keyhole in the BM, as shown in Figure 2.6d.

2.6 ADVANTAGES AND DISADVANTAGES OF THE FSW PROCESS

2.6.1 Advantages

In FSW, shielding environment and filler material is usually not essential for joining. Thus, it does not produce any harmful emissions and is also safe for operators, as shown in Figure 2.1. It is, therefore, considered to be an energy-efficient, environment-friendly, and a clean material joining process. The process is completely mechanical and, therefore, the welding operation and weld energy input are accurately controlled. The various advantages of FSW process over conventional fusion welding processes are shown in Figure 2.7.

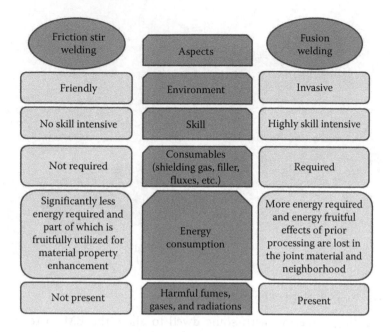

FIGURE 2.7   Advantages of FSW over fusion welding.

Some of the advantages of FSW are listed below:

- Better mechanical properties in the as-welded condition

- Obtaining weld with maintained original properties

- Material that could not be welded earlier can be successfully welded

- Improved safety due to the absence of toxic fumes

- No filler or gas shield is required for reactive material

- Less-skilled operator is required

- No limitation for joint configuration and welding position, that is, it can operate in all positions (horizontal, vertical, overhead, etc.) due to the absence of weld pool

- Uniform weld bead with good appearance, thus reducing the machining after welding

- Low environmental impact

- Welding preparation not usually required

- Smaller HAZ that improves the joint quality for heat-treatable alloys

- Low energy consumption

2.6.2 Disadvantages of the FSW Process

- Low welding speed compared to some automatic fusion welding processes

- High forces are required for rigid clamping of BM

- Key hole is left at the end of the weld

- For avoiding key hole in the welded plate, run-on/run-off plates are required

- Robust fixture is required for counteracting the large plunging forces

- High investment is needed

- Less flexible compared to manual and arc welding processes

- Slower production rate than some fusion welding techniques

- Different size of the tool pin is required when welding materials of varying thickness

To overcome some of the drawbacks of FSW, an engineer at Marshall Space Flight Center (MSFC), NASA, helped to design a computer-controlled automatic retractable pin tool, as shown in Figure 2.8. This tool automatically retracts the pin inside the tool shoulder after the completion of weld before withdrawal from the welded plates, which leads to a smooth hole closure at the end of the weld. This design also allows adjustable pin length for varying thickness and prevents the use of different pins of varying length for different plate thickness.

2.7 APPLICATIONS OF FSW

Application of FSW in several industrial sectors resulted in remarkable benefits in terms of performance and cost reduction. FSW possesses numerous applications in different industries including aerospace, shipbuilding, railways, automobile, etc. Dissimilar material joining is becoming increasingly important to fulfill the requirements of engineering structures and components with reduced weight, improved strength, and functionality. FSW has been adopted extensively for welding aluminum

FIGURE 2.8   Retractable pin tool. (Courtesy of NASA Marshall Space Flight Centre.)

alloys in various industries owing to economic and environmental benefits. Dissimilar material joining such as aluminum to aluminum, aluminum to steel, and aluminum to copper significantly utilizes the benefits of the properties of both the materials. For instance, joining aluminum alloy and steel is of significant importance, and recently, Honda has successfully performed such dissimilar welding in a vehicle suspension system for large-scale production (Honda, 2012).

Formation of brittle intermetallic compound (IMC), difference in softening temperatures, and thermal expansion act as barriers in dissimilar materials joining, which adversely affects the performance of components and structures. FSW results in reduction in formation of IMCs during dissimilar material welding, which in turn improves the performance of components and dissimilar structures for industry-oriented applications. Sound joining of dissimilar aluminum alloys by FSW enhances the use of high strength-to-weight ratio aluminum alloys in aerospace, automobile, and railways. It makes mass production of light-weight transportation system that results in remarkable reduction in fuel consumption. For more than a decade, FSW has been adopted by the aerospace and automotive industry to join different aluminum alloys. The list of companies that adopted FSW as a joining process in several applications is presented in Table 2.1.

TABLE 2.1   Sample List of Companies Adopting FSW in Applications in the Last Two Decades

Year	Application	Company
1995	Heat exchangers	Marine Aluminum, Norway
1996	Shipbuilding	Marine Aluminum, Norway
1998	Delta II rockets	Boeing, United States
1999	Shipbuilding	SAPA, Sweden
2000	Automotive components	SAPA, Sweden
2000	Laser system housings	General Tool, United States
2001	Motor housings	Hydro Aluminum (formerly Marine Aluminum), Norway
2001	Suburban trains	Alstom, Germany, and Hydro Marine
2001	Automotive components	Showa, Japan
2001	Train bodies	Hitachi, Japan
2002	Automotive components	Tower Automotive, United States
2003	Aircraft structure	Eclipse, United States
2003	Commercial shipbuilding	Advanced Joining Technologies, United States
2004	Space shuttle external tank	Lockheed Martin, United States
2006	Shipbuilding	Friction Stir Link, United States
2012	Automotive components	Honda Motor Company, Japan

Source: Reprinted from *Advances in Friction Stir Welding and Processing,* A. Amini, P. Asadi, P. Zolghadr, Friction stir welding applications in industry. Woodhead Publishing Limited, Cambridge, UK 671–722, Copyright 2014, with permission from Elsevier.

2.7.1 Shipbuilding and Marine Industries

The shipbuilding and marine industries adopted the FSW process for joining 5xxx and 6xxx series aluminum alloys. The process is suitable for the following applications:

- Panels for decks, sides, bulkheads, and floors
- Aluminum hulls
- Boat internal surface
- Superstructures
- Helicopter landing platforms
- Ship body structures

2.7.2 Aerospace Industry

The US aerospace industry uses FSW for fabricating the satellite launch vehicle's tanks made of high-strength heat-treatable aluminum alloys.

The first rocket fabricated by FSW was successfully launched in 1999. Components and structures made of different materials are used in aerospace industry as they undergo different loading conditions and require different properties in a single component. FSW is successfully employed in military aircraft for welding of stringers and ribs with skins. This results in weight saving and reduction in costs associated with manufacturing compared with riveting and other mechanical fastening. Aircraft fuel tanks made of AA2219 are commercially joined by using FSW. The process has been applied by various companies such as Boeing, Lockheed Martin, Air Bus, etc. This method has also been used in many of the world's space launch vehicles, including the Space Shuttle, Delta II and IV, SpaceX Falcon 9, and Ariane.

FSW is used in the fabrication of various parts and components of aircraft and space vehicles such as wings, fuel tanks, stringers, etc. Further, it is also used in the repair of faulty welds.

2.7.3 Railway Industry

FSW is widely used in production of rail car bodies made of aluminum alloys such as Hitachi superfast trains (Shinkansen), which can reach speeds up to 320 km/h.

Applications include the following:

- Bodies of high-speed trains made of high-strength aluminum alloys
- Underground carriages and trams
- Railway tankers and goods wagons

2.7.4 Automobile

The use of aluminum tackled two largest challenges faced by the automobile industry. First is vehicle's weight reduction and second is passengers' safety. Aluminum alloys provide enough strength and is also light in weight. The problem associated with aluminum is its welding. Conventional fusion welding processes rarely achieve sound welding of aluminum alloys. Following the invention of FSW, aluminum alloys are increasingly being used in vehicle bodies. Currently, FSW is widely used by various car manufacturers worldwide such as Honda, Mazda, Ford, Audi, and Tower Automotive. Tower Automotive substituted the gas metal arc welding (GMAW) by FSW because of reduction in weight by 40% and improvement in mechanical properties by two times compared to GMAW for 6000 aluminum alloy series (ESAB, 2012).

In recent years, automobile manufacturers have shown interest in joining various automotive parts by FSW and investigation is going on for its commercial application.

Some of the potential applications of FSW in automobile industry includes joining of

- Front portion of engine
- Alloyed wheel rims
- Fuel tankers
- Bodies of heavy duty vehicles
- Tailored blanks

2.7.5 Construction Industry

The use of FSW in construction industry needs its portability and portable FSW machines have been introduced for making parts and components including the following:

- Facade panels made from aluminum, copper, or titanium
- Window frames
- Aluminum bridges
- Reactors for power plants
- Pipe fabrication

2.7.6 Electrical Industry

FSW is widely employed by electrical industry in

- Electric motor housings
- Encapsulation of electronics
- Electrical connectors

2.8 COMMERCIALIZATION OF FSW

2.8.1 Shipbuilding

FSW is widely used in Japan for the production of honeycomb and seawater resistant panels. Combined passenger and freight ship named "Super

Liner Ogasawara" was built by the Mitsui Engineering and Shipbuilding (MES) by using FSW. In 2004, New Zealand and Australian navy required 55-m-long Inshore Patrol Vessels. Significant portion of such vessels was specified by the naval architect to be built by using FSW. The Donovan Group implemented FSW in mid-2005 for manufacturing Inshore Patrol Vessels in Whangarei (NZ). They modified and retrofitted a large CNC gantry milling machine capable of performing FSW for mass production of such panels. These panels were used in superstructures of Inshore Patrol Vessels.

Scandinavian aluminum extruders employed FSW process commercially in 1995 for the first time to manufacture deep freezer panels made of hollow aluminum alloy. Wide aluminum panels for cruise ships and offshore oil platforms are commercially being produced by FSW. Large number of panels with hundreds of kilometers weld length has been produced and used in "The World" cruise ship in Haugesund by Marine Aluminum since 1996. The commercial application of FSW in fishing boats for joining hollow aluminum extruded panels for deep freezing of fish was started in November 1996 at Sapa in Finspang (Sweden).

2.8.2 Aerospace

FSW offers various advantages over fusion welding processes, which has been commercially derived by the aerospace industry. Boeing is among the first company in the United States that commercialized FSW and started using it to fabricate the fuel tanks for space program. Significant reduction in welding cost was achieved by Boeing after employing FSW in place of tungsten inert gas (TIG) welding in the fabrication of rocket fuel tanks. For instance, Boeing reported that the use of FSW in Delta II and Delta IV resulted in 60% cost savings, and reduction in the manufacturing time from 23 to 6 days. FSW was successfully employed for making 2100 m weld in Delta II rocket and 1200 m welds in Delta IV by July 2001.

The tank domes and walls of Falcon 9 are made by FS-welded aluminum lithium alloy AA2198, and the Falcon 9 tank is claimed to be the largest fully FS-welded structure in the world. The 55-m-long vehicle (Falcon 9) is assembled horizontally and rolls out to the pad for being placed vertically for launch (Figure 2.9).

Eclipse Aviation Corporation, New Mexico, replaced traditional riveting and other mechanical fastening processes by FSW in their commercial aircraft production to lower assembly time and cost (ESAB, 2012). They succeeded in eliminating 7000 rivets and other fasteners in their

FIGURE 2.9   Circumferential and longitudinal friction stir welds in a Falcon 9 tank. (Courtesy of SpaceX.)

Eclipse 500 by implementing FSW. The use of FSW in Eclipse 500 aircraft improved the strength and fatigue life as compared to the conventional fastening techniques. Airbus, aircraft manufacturing company, also showed interest in using FSW in their aircrafts. The first application of FS-welded joints by Airbus was seen in the welded floor panels of A400M military transport aircraft. The use of FSW minimized aircraft weight and enhanced its mechanical properties.

The first commercially successful application of FSW in aerospace industry was done in the form of the Space Shuttle's gigantic External Tank, which is the largest element of the spacecraft (Space Shuttle Technology Summary, 2001). FSW was used to join Aluminum alloy 2219 for commercialized production of external tanks as shown in Figure 2.10.

NASA's Michoud Assembly Facility joined bulkhead and nosecone of the Orion spacecraft using FSW (Figure 2.11). Fabrication of space vehicles using FSW resulted in improved strength and reduced weight. Orion spacecraft is intended to carry crew and facilitate human exploration. FSW is used in Orion spacecraft with the aim of achieving defect-free and high-strength structures that are desirable for space travel.

NASA's space launch system (SLS) uses FSW to bond the core stage's rings, domes, and barrel segments and found defect-free joints with higher strength compared to the conventional joining methods. Friction stir-welded SLS Launch Vehicle Stage Adapter structural test article at NASA's

FIGURE 2.10 Space shuttle and its gigantic external tank (Space Shuttle Technology Summary, 2001). (Courtesy of NASA.)

MSFC in Huntsville, Alabama, is shown in Figure 2.12. It connects the SLS core stage to the interim cryogenic propulsion stage.

2.8.3 Railways

Large panels of aluminum extrusions are used in the manufacturing of trains in Japan. Trains are being assembled by using FSW from aluminum

FIGURE 2.11 FS-welded bulkhead and nosecone of the Orion spacecraft at NASA's Michoud Assembly Facility. (Courtesy of NASA.)

FIGURE 2.12 SLS launch vehicle stage adapter at NASA's MSFC. (Courtesy of NASA.)

extrusions. Development in the tool design and joint configuration makes the FSW an effective method for joining aluminum panels in train manufacturing. Modern railway carriages are fabricated from longitudinal aluminum extrusions by using FSW. Improvement in the performance of FS-welded joints revolutionized the process of train building by railway vehicle industry. Combination of good weld performance and cost effectiveness enabled the use of FSW in the manufacturing of trains and trams. FSW is being used in the train side walls since 2001 and train floor panels since 2002 in Munich suburban trains made by Sapa. It was used to weld up to 23-mm-thick aluminum plates in an acceptable time span and it replaced the mechanized metal inert gas (MIG) welding used for making underframe area of rolling stock.

Low distortions in FS-welded parts enabled the use of FSW in joining of aluminum extrusions in double skin designed coach by Hitachi in Japan. It eliminates the cost of filing and straightening of arc welded thin aluminum parts. Hitachi started exporting FSW trains to Europe for domestic services. FSW has been used by Nippon Sharyo for welding floor panels of new Shinkansen. Nippon Light Metals have also made use of FSW for subway rolling stock. They reported that over 3 km of welds had been produced by 1998.

Friction stir spot welding is being used by Kawasaki Heavy Industries (Japan) for attaching stringers to roof panels. They used FSW for assembling of aluminum car body shell. Low heat input produced in FSW resulted in improvement in the flatness and visual appearance of the skin panels of the car.

2.8.4 Automobile

In 1998, TWI started using FSW for joining aluminum tailored blanks for door panels in car bodies and demonstrated new concepts on FSW in a project involving various car manufacturers such as BMW, EWI, Rover, Ford, General Motors, and Volvo. Encouraging performance of the FS-welded joints in car door panels enabled the use of FSW in other automotive components and since then FSW and its variant Friction Stir Spot Welding (FSSW) are being used in the mass production of aluminum automotive components and dissimilar aluminum with steel automotive parts throughout the world (Thomas et al., 2006).

Honda Motor Corporation has implemented FSW for joining aluminum alloy and steel in front structural component of Honda Accord production. The front subframe that carries some suspension components and the engine is made of press formed steel halves and die cast aluminum. FSW was applied at various locations to weld the lap configuration of aluminum to the steel. As claimed by Honda, 25% reduction in the total body weight was achieved compared to previous conventional steel subframe along with 50% reduction in electrical consumption (Honda, 2012).

Mazda Motor Corporation has developed direct friction stir spot joining technology for dissimilar welding of aluminum alloy with steel, and applied it to join the trunk lid of the Mazda MX-5. Figure 2.13 shows Mazda's dissimilar friction stir-welded deck lid with aluminum sheet to galvanized steel brackets. In addition to the prominent weight reduction, Mazda also claims that spot welding process improves the potential of coupling aluminum parts to steel in vehicle bodies and also helps in minimizing production cost.

Ford used FSW as a fabrication process for the first time in June 2004, in its Ford GT for welding multipiece central aluminum. Ford reported improvement in dimensional accuracy and 30% enhancement in joint strength by using FSW in comparison to the automated GMAW. FSW

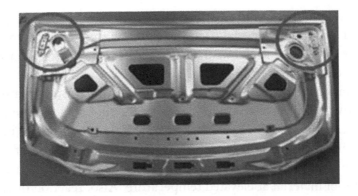

FIGURE 2.13   FSW of aluminum with the galvanized steel in MX-5 trunk lid and bolt retainer by Mazda (2005).

finds itself as a feasible joining technology for welding different aluminum alloys in the last 15 years and has been efficiently applied on large scale for joining parts and components in various industrial sectors (Kallee, 2010).

2.9 FSW TOOL MATERIAL

Tool material plays a vital role in determining suitability of a tool for welding a particular material. Generally, during FSW temperature reaches to around 85% of the workpiece material's melting point; such a high operating temperature requires tool of high hardness, high hot strength, hot hardness and good wear resistance, and high-temperature chemical stability. Such properties of tool material prevent its deformations, twisting, breakage, and erosion during FSW (Rowe and Thomas, 2005). Tool material also affects the temperature attained at the SZ of welded joint. Adequate generation and distribution of heat at the interface of tool and workpiece during FSW depends on coefficient of friction between tool and workpiece and thermal conductivity of tool and workpiece. Erosion of tool by sticking and slipping of the workpiece on the tool depends on hot hardness of the tool material. A general consideration is that the hardness of tool material should be two or three times higher than that of the workpiece material. Properties such as friction coefficient, thermal conductivity, hot hardness, and toughness of the tool depend on the tool material. Friction coefficient governs the sticking and slipping of the material around the tool pin, which affects material movement and heat generation. Inappropriate coefficient of friction results in improper material movement and insufficient heat generation leading to degradation of joint quality. The rate of heat dissipation during welding is significantly affected by thermal conductivity of the tool and workpiece materials, which in turn governs the thermal field and flow stress around the weld. High thermal conductivity of tool dissipates excessive heat from interface of workpiece and tool that requires high rotational speed and low welding speed, which in turn minimize productivity. This at the same time keeps the tool cold and minimizes tool degradation as well. Low thermal conductivity minimizes heat loss that results in fruitful utilization of generated heat. Erosion of the tool due to rubbing with the material being welded is controlled by the hardness of tool material at elevated temperature. Also, reaction of ambient oxygen with hot tool material enhances the degree of erosion of tool. Tool's reaction with oxygen is an important consideration for selection of the tool material. Adequate values of friction coefficient, thermal conductivity, and hardness of the tool depend on the material of workpiece, process variables, and tool geometry.

Tool steel is generally used for welding low melting point material such as aluminum alloys, copper alloys, and magnesium alloys for up to 700°C of working temperature. Various tool steel materials such as H13, high carbon high chromium (HCHCr), SKD61, high carbon (HC), high-speed steel (HSS), etc. are successfully used for welding of dissimilar aluminum alloys. Tool material used for FSW of dissimilar aluminum alloys along with their geometry is presented in Table 2.2.

Tools used for welding high-strength materials such as titanium alloy, nickel alloys, steel, etc. are polycrystalline cubic boron nitride (pcBN), nickel, and cobalt-based super alloys. In addition, a few specific grades of tungsten carbide are also being used. The characteristics of the tool material are crucial for FSW and the desired characteristics depend on workpiece material to be welded. High hardness of tool is desirable for welding wide range of materials (Zhang et al., 2012). PCBN, being a super hard material, is most suitable for FSW of high-strength material, such as titanium and steel, due to its good thermal and mechanical performance, but the problem is associated with its machinability, which makes forming of FSW tool geometry very difficult. Machining of such tools and also those made from tungsten carbide is very difficult especially in case of complex pin geometry. Development of tools having desirable properties, ease of manufacture, and low cost has made the welding of dissimilar aluminum alloys easy and cost effective. Various tool materials that are available for FSW are presented in Table 2.3.

TABLE 2.2   Tool Materials and Geometries Used for FSW of Several Dissimilar Aluminum Alloys

Workpiece Material	Tool Material	Tool Geometry	References
2024-T3 and 7075-T6 3 mm	SKD61	SD: 12, PD: 4, PL: 0.6, PS: cylindrical threaded	Khodir and Shibayanagi (2007)
2024-T3 and 6082-T6 sheets 0.8 mm	56NiCrMoV7-KU (Italy grade)	SD: 6, PD: 1.7, PL: 0.6, PS: cylindrical	Scialpi et al. (2008)
AA6082 and AA2024 4 mm	Threaded C40 steel tool	SD: 9.5, PD: 3.8, PL: ----, PS: threaded conical	Cavaliere et al. (2009)
AA5052 and AA6061 5 mm	HSS tool	PL: 4.8 mm, PD: 6 mm, SD: 25 mm, PS: cylindrical	Kumbhar and Bhanumurthy (2012)
AA6351-T6 and AA5083-H111 6 mm	HCHCr steel	SD: 18, PD: 6, PL: 5.7, PS: straight square	Palanivel et al. (2014)
AA7075-T6 and AA6082-T6 8 mm	H13 steel	SD: 23, PD: 5, PL: 7.9, PS: triangular frustum	Aval (2015)
AA2219 and AA 7039 6 mm	SS 310	SD: 19, SS: flat, concave, PD: 7, PL: 5.8, PS: threaded cylindrical	Venkateswarlu et al. (2015)
AA5083 and AA6063 4.75 mm	Tungsten carbide	SD: 20, SS: flat, PD: -----, PL: 4.4, PS: tapered cylindrical	Khan et al. (2015b)
AA7475-T761 and AA2219-O 2.5 mm	HCHCr steel	SD: 20, SS: flat, PD: 6, PL: -----, PS: threaded cylindrical	Khan et al. (2017)

SD, shoulder diameter (mm); SS, shoulder shape; PD, pin diameter (mm); SD, shoulder diameter (mm); PL, pin length (mm); and PS, pin shape.

TABLE 2.3   Characteristics of FSW Tool Materials for Butt Welding

Tool Material	Advantages	Disadvantages
H13	Easy machinability, good elevated temp strength	Severe tool wear for high-strength materials or metal matrix composites (MMC)
SKD61	Good thermal fatigue resistance	Tool wear with complex pin profiles
HCHCr	High hot hardness compared to other tool steel	Difficult to machine in hardened condition
Tungsten	High hot hardness and strength. Suitable for high-strength materials.	Poor machinability, expensive, low coefficient of friction with aluminum

Various properties that should be possessed by a suitable tool material for performing FSW are listed below:

• Good strength and wear resistance at elevated temperature for bearing high plunging forces

• Good dimensional stability for repeated use

• Good coefficient of friction with the workpiece for adequate generation of frictional heat

• Nonreactive with oxygen and workpiece material

• Good machinability for making complex geometry

• Hot hardness should be high enough to complete long weld

• Cost effectiveness

2.10 FSW TOOL DESIGN

The two important characteristics of FSW tool geometry are as follows: (1) its shape should be as simple as possible to reduce the cost and time and (2) it should be able to produce adequate stirring effect to produce sufficient material movement. Tool design is a critical factor during FSW and plays a vital role in plasticized material flow and localized heating at the SZ that significantly affects defect formation and postweld mechanical properties. FSW tools are designed to generate sufficient heat due to friction between tool and BM and plastic deformation of BM and also to obtain effective material flow. Tool comprises two main portions: (i) tool shoulder and (ii) tool pin, as shown in Figure 2.14. Tool shoulder generates heat due to friction, which is necessary for softening the BM being

FIGURE 2.14   FSW tool.

welded and forges the material being stirred behind the tool pin in order to properly consolidate the stirred material. It also restricts the material from extruding outside the shoulder. Tool pin generates heat by plastic deformation of the BM and promotes its movement around the pin (Kim et al., 2010). Different designs of tool shoulder surface that are being used in FSW include flat, concave, scrolled, concentric circle, etc. Scrolled and concentric shoulder design is shown in Figure 2.15. Tool shoulder surface plays an important role in stirring of the material as well. Different action is served by various shoulder surfaces features depending on the requirement. Flat shoulders are easy to make and simple in design, and can be used for welding aluminum alloys except in special cases where enhanced

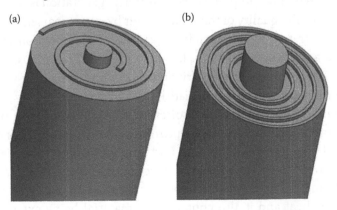

FIGURE 2.15   FSW tool shoulder designs: (a) scroll and (b) concentric.

stirring action and material consolidation are required. Concave shoulders are designed to restrict the stirred material within shoulder to minimize flash formation. Concave tool shoulder traps the flowing material under tool shoulder, which results in better consolidation of flowing material. Scrolled and concentric-type shoulders are designed to enhance degree of material stirring for proper mixing of the BMs to improve joint properties. Flat end surface having spiral channel toward the center constitutes a scroll shoulder. Spiral channel directs the flowing material from end toward the pin to prevent expelling of material outside the shoulder. Scroll shoulder eliminates the use of high tool tilt for consolidating the material behind pin as it works normal to the workpiece, which reduces thinning of weld region. Spiral groove in scrolled shoulder promotes plastic deformation and frictional heat. Scroll shoulder faces problems during welding of materials with different thickness. Along with the design of tool shoulder features, its diameter also significantly affects the process in terms of heat generation and material movement.

2.10.1 Shoulder Diameter

In FSW, heat is generated by friction and plastic deformation of the material and shoulder produces heat by frictional sticking and sliding of the material under it. Large shoulder diameter generates more friction and consequently increases heat input. Sticking of material is responsible for material flow in the stirred region. Tool shoulder diameter governs heat generation by friction between tool shoulder and work piece as it increases the contact area. Also, large shoulder diameter affects large portion of workpiece, which results in increased HAZ leading to deterioration of the postweld mechanical properties. Excessive heat generation is not desirable as it degrades the quality of the welded joint by grain coarsening and dissolution of strengthening precipitate (in case of precipitation hardening materials) and loss of cold work. It is also observed that the smaller tool shoulder diameter leads to less frictional heat and hence the weld metal consolidation may not be appropriate in the SZ. It consequently affects the joint strength due to insufficient softening of material causing improper material mixing in SZ. Evidently, optimum shoulder diameter is required that could generate sufficient heat and promote proper material movement, minimize torque, and traverse force and produce small HAZ.

The diameter of the tool is important because the shoulder generates most of the heat and it also contains the material being extruded from AS. Further, tool shoulder forges the flowing material behind the pin preventing plasticized material from escaping out of weld region and consequently ensuring proper consolidation of joint. It has been established through several findings that the shoulder diameter should neither be too small nor too large. For example, Padmanaban and Balasubramanian (2009) in their investigation considered three shoulder diameters (15, 18, and 21 mm) and demonstrated that the joint fabricated using 18 mm diameter yielded defect-free and fine-grained nugget region, which subsequently showed higher hardness and superior tensile properties. Similar observations were made by Elangovan and Balasubramanian (2008) during FSW of AA6061 aluminum alloy. Thus, shoulder diameter should be carefully chosen based on the required heat input.

2.10.2 Pin Geometry

Tool pin geometry includes shape and size of pin. Size of the pin is related to shoulder diameter as their relative size affects the heat generation and plastic deformation. Pin profile mainly governs the material movement. Flow of plasticized material is significantly influenced by type of tool pin profile. During plunging of tool into faying surface, the heat is primarily generated by friction and plastic deformation of material at pin–workpiece interface, which softens the material in the vicinity of pin and moves the same around it by sticking action. FSW process uses combination of extrusion and forging processes for joining the materials. Pin extrudes the softened material around it from leading to the trailing side and shoulder forges the flowing material behind the pin for its consolidation.

Various pin profiles are used in FSW tool such as, cylindrical and tapered cylindrical (i.e., frustum of a cone, square, hexagon, threaded cylindrical, threaded cam, and threaded tri flute as shown in Figure 2.16). Each profile performs different functions in terms of material movement and material mixing. Cylindrical and tapered cylindrical pin moves material around the pin, whereas threaded pin moves it around the pin as well as in vertical direction. Square, hexagonal, and cam pin profiles produce pulsating action due to flat faces during stirring of flowing material. Good mixing is obtained by the pulsating action of the pin, which results in improved joint strength. At the same time, pin profiles other than cylindrical tends to wear out in due course of time leading to a conical form. On prolonged usage, profiled pins wear out and acquire tapered cylindrical or cylindrical shape. Hence, for long run and large number of runs with the same pin, optimum pin profile is used that can sustain its shape as well as provide better stirring action.

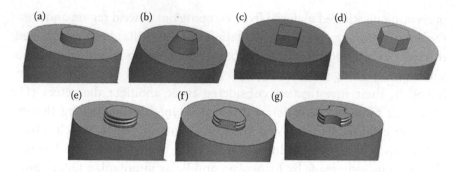

FIGURE 2.16 Different FSW tool pin profiles: (a) cylindrical, (b) tapered cylindrical (i.e., frustum of a cone), (c) square, (d) hexagon, (e) threaded cylindrical, (f) threaded cam, and (g) threaded tri flute.

2.11 FSW PROCESS PARAMETERS

Precise criterion for selection of FSW process parameters is still evolving and as of now it is based on experience and outcomes of experimental trials. Often the parameters produce contradicting effect on responses; consequently, selection of process parameters should be based on combined effect in such a way that defect-free joints with good joint efficiency and high production rate are obtained. Important FSW parameters include tool travel speed, tool rotational speed, tool tilt angle, pin offset, and plunge depth. These parameters also influence heat generated during FSW process. There are some other factors related to tool that also influence resulting weld such as pin features, tool material, diameter of tool, and type of tool shoulder. Evolution of some responses such as peak temperature, longitudinal force, torque, and power requirements are determined by FSW process parameters and type of BM.

2.11.1 Rotational and Traverse Speeds

Tool rotation, that is, rpm helps in material movement around pin and mixes the material to complete the joint. Increase in the rotational speed increases friction heat, which in turn increases temperature in SZ. There is nonlinear relation between frictional heat and rotational speed as the interface frictional coefficient changes with increase in tool rotation and temperature. Typically, a study has revealed that an increase of rotational speed from 300 to 650 rpm increased peak temperature of SZ by 40°C and a further increase from 650 to 1000 rpm increased the peak temperature by 20°C (Tang et al., 1998). This result, like several others, suggests that the temperature increases with slower rate at higher rotational speed. Increase in tool rotational speed significantly increases the peak temperature and decreases the torque. Lower rotational speed generates insufficient heat, which results in inadequate plasticization of material leading to lower joint strength. Higher rotational speed generates more heat, which may often cause undesirable metallurgical transformation such as, dissolution and coarsening of strengthening precipitates, grain growth, and minimizing dislocation density, which will degrade the joint strength. Therefore, optimum rotational speed must be selected for defect-free joints with good postwelding mechanical properties.

Tool traverse speed affects distribution of generated heat along length of weld during welding. Increase in tool traverse speed, also known as welding speed, decreases the peak temperature in SZ. However, higher welding speed reduces heat input at weld region, which makes the material flow difficult and thereby may result in slight increase in torque. Tool traversing in welding direction moves the plasticized material from AS to the back of tool pin to complete joining. Higher welding speed may lead to large longitudinal forces, which may also result in tool wear and in extreme cases cause tool breakage. Also, high welding speed is one of the causes of formation of tunneling defect at the bottom of SZ and TMAZ interface in AS due to improper consolidation of flowing material behind pin. At higher welding speed, tool moves ahead before depositing sufficient material necessary to fill space behind it vacated by tool creating a continuous void, which may be termed as tunnel. The size of defects increases with increase in the welding speed (Crawford et al., 2006).

During FSW, a material to be welded undergoes SPD due to stirring and simultaneous action of heat and SPD causes dynamic recrystallization, which results in grain refinement. Grain refinement is affected by degree of plastic deformation and amount of heat generated at SZ. Indirect effect is that the size of recrystallized grains increases with increasing rotational speed and decreasing tool traverse speed. Tool rotation and traversing speed must be selected to ensure efficient welding. Lower rotational speed combined with higher traversing speed results in colder weld whereas, higher rotational speed along with lower traversing speed results in hotter weld.

In FSW, it is essential to decide (i) how fast the tool should be rotated and (ii) how quickly it should be allowed to traverse along the interface rotates. Tool rotational and traversing speed must be carefully selected to make the flowing material hot enough for undergoing the required plastic deformation and reduce the forces experienced by the tool. At the same time, the flowing material must not be overheated to cause degradation of mechanical properties of the joint. Therefore, a process window of the parameters is required to produce defect-free welds by avoiding too cold and too hot weld.

2.11.2 Tool Tilt Angle

Inclination of the tool axis from the normal of the workpiece surface is referred to as tool tilt angle. It is defined as the angle between the tool axis and normal line to the workpiece surface, as shown in Figure 2.17. It significantly affects the rate of heat generation, material movement, and consolidation of flowing material behind the tool pin. Axial force applied by the tool and peak temperature increases with tool tilt angle. Tool tilt angle helps in preventing the flowing material from being expelled. Also, tool tilt angle affects the shape of SZ, which indicates its effect on vertical and horizontal flow of material in SZ. Tilting the tool by some degree in such a way that the front of tool is higher than its rear, helps in adequate forging of the plasticized material behind the tool to complete the joint.

2.11.3 Plunge Depth

Plunge depth is defined as the depth of lowest point of shoulder below the surface of welded plate, as shown in Figure 2.17. It is found to be a critical parameter for ensuring weld quality. Plunge depth is an important

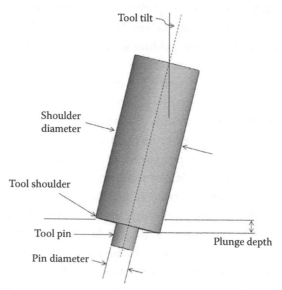

FIGURE 2.17 FSW tool with plunge depth and tilt angle.

process parameter of FSW that plays a vital role in heat generation during welding and, it also controls the forging of flowing material. Increase in plunge depth increases axial force, which in turn increases frictional heat at tool–workpiece interface. Further, increase in plunge depth results in higher heat input, which leads to grain growth and formation of IMCs that directly affect strength and ductility of the joint. On the other hand, low plunge depth results in lower heat generation, which causes insufficient plasticization of material leading to inadequate material mixing that may lead to the formation of defects. Also, lower plunge depth decreases peak temperature that causes increase in flow stress leading to slugging material movement. Therefore, appropriate plunge depth is important for producing good-quality joints by ensuring adequate forging pressure required to consolidate flowing material properly as well as full penetration of the tool inside BM.

2.11.4 Tool Pin Offset

A shift of tool axis on either side of the joint line is regarded as tool pin offset or simply tool offset. It has a significant effect on the distribution of generated heat on either AS or RS, and also on the mixing of two materials during FSW of dissimilar materials. In addition, pin offset influences postweld mechanical properties due to differences in physical, chemical, and mechanical properties in case of welding dissimilar materials. Previous studies showed that for FSW of dissimilar materials, poor joints were obtained at weld line of butted plates and the pin offset toward softer materials was usually adopted to achieve the defect-free joints (Khan et al., 2015b). Cole et al. (2014) reported that higher joint strength is obtained with tool offset toward stronger material. Joint properties can be improved by offsetting the tool toward either side of the joint line based on the materials to be welded.

2.12 FSW EXPERIMENTAL SETUP

FSW process utilizes heat due to friction and plastic deformation to perform welding operation. Large forces and torque are required to generate frictional heat and also for plastic deformation. A heavy duty robust experimental setup is required so as to withstand forces and torque developed during welding. Various companies provide complete setup to perform FSW of different materials in various joint configurations. Such machines are too expensive, which is the major obstacle to perform research in the area of FSW. An experimental setup that is capable of performing FSW in a cost-effective manner is required and at the same time, it should be robust to bear the loads generated during welding.

The use of a heavy duty vertical milling machine with sufficient power is suitable to perform FSW of different materials. A standard vertical milling machine is good for metal-cutting application and does not have interface to accept FSW tool and mount plates for FSW. Thus, an appropriate tool adopter and work fixture need to be devised in order to enable this machine to perform FSW. A tool adapter is required to clamp the tool effectively so that there is no slipping of the tool under heavy torque generated during FSW and also to withstand high temperature generated at the tool. A robust work fixture is also required to withstand high plunging force applied by the tool during plunging into BM and traversing force that are involved when the tool starts to traverse after plunging. The fixture should also serve as a heat sink for high temperature generated at the SZ.

To perform FSW operation, tool is clamped in the collet carried in FSW tool adaptor, which in turn is mounted on the machine spindle adapter. BMs to be welded are clamped rigidly on the work fixture, which is mounted on the machine table having three operational directions. After proper clamping of the tool and BM, tool is rotated and work table moves upward until the tool plunges up to the required depth inside the BM. After attaining the required height by the work table plunging is stopped and movement is given to it in traverse/welding direction to complete the joining process.

Various other machines are also required for preparing welding specimen/test coupons for testing the quality of welded joints. To perform the FSW and testing of dissimilar aluminum alloy-welded samples, following machines/equipments are required.

- Shaper machine for dimensioning the workpiece
- Surface grinder for making weld specimen with uniform thickness
- Tool and cutter grinder for preparing the FSW tools
- A heavy duty vertical milling machine adapted for FSW
- FSW fixture
- Wire E.D.M for making sample for testing
- Tensometer for tensile testing

- Charpy impact testing machine for impact test

- Metallurgical polishing machine for grinding and polishing

- Optical microscopes for microstructure

- Stereozoom microscope for macrograph

- Scanning electron microscope (SEM) for microstructure, particle composition, and fractography

- Transmission electron microscope for precipitate evolution

- Microhardness testing machines for measuring microhardness

2.12.1 FSW Machine

FSW process requires either a dedicated FSW machine or robust vertical milling machine capable of performing FSW. FSW machine is very cost intensive at commercial scale and heavy duty vertical milling machine requires special arrangement for performing FSW. Special work fixture is required for performing FSW on conventional vertical milling machine. Robust vertical milling machine with high spindle power can be used to perform FSW. A conventional vertical milling machine may be adapted to perform FSW where friction stir-welded joints may be made on various BMs. Special tool adopter is required to enable vertical milling machine to perform FSW.

FSW machines supplied by ESAB are being used by several companies such as Boeing, Tower automotive, Marine Aluminum, SAPA, etc. for producing FS-welded components. ESAB is one of the largest FSW equipment manufacturers around the world. ESAB supplied many robust and accurate FSW machines for mass production.

A typical vertical milling machine retrofitted to perform FSW is shown in Figure 2.18. The vertical milling machine can be retrofitted to an FSW machine with the help of a suitable tool adopter and work fixture. Machine must possess a heavy power motor to provide the torque required during welding.

2.12.2 FSW Work Fixture

FSW can be performed on high-power conventional vertical milling machine with suitable tool adopter and work fixture. The welds are created by the combination of frictional heating between tool and BM and plastic deformation of the BM. In FSW, the rotating tool plunges into the abutting

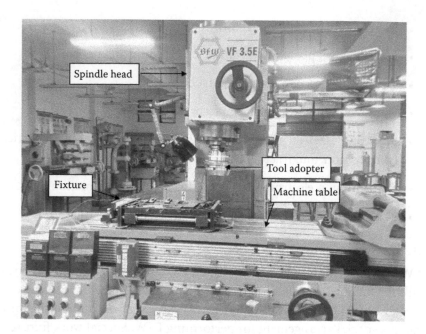

FIGURE 2.18 FSW machine (retrofitted).

surface of the BM and traverse into the welding line to produce the welded joint. Friction heat generated between the tool shoulder and BM causes softening of the metal along the joint without reaching the melting point of the metal. Large vertical force acts during plunging of the tool into the BM and longitudinal force acts during traversing of the tool along the joint line. To withstand the forces, the BM should be rigidly clamped on the robust fixture to avoid damage of the machine tool and prevent improper welding. To avoid this, work fixture with robust anvil and large number of clamps are used to hold the BM rigidly. The idea of using large number of clamps has been limited due to inflexibility of the BM dimension. This idea has yet not been extended to allow BM, with variable dimension, simultaneously hold the BM from top, transverse side, and longitudinal side. Robust anvil with optimum number of clamps is required to provide adequate clamping to the BMs so as to prevent them from vibrations and bending due to heavy forces generated during welding.

The work-holding system should not only properly clamp the joint being welded, but also be able to rigidly clamp BMs of variable dimensions so that it can be utilized for a number of dimensionally varied BMs. In addition, the time required to clamp and declamp for changing the BM being welded should be minimum. It should provide sufficient clamping force to make the process robust, which in turn will result in energy saving by minimizing the energy that goes waste in making defective joints due to improper clamping of the BMs.

The major issue in designing of FSW fixture is to securely clamp the BMs during welding process. The plunging of rotating tool into BM and the high forces evolved during welding tend to separate the BMs. The moving tool pushes BM away from each other due to which defects such as tunneling defect may be generated if BMs are not rigidly clamped from the side. The heat generated in the process often causes bending or warping of the edges of the BMs. Also, there should be a sufficient distance between the moving tool and the clamps (used for holding BMs) in order to avoid collision. Further, in case the clamps occupying more area of the BM, it may lead to the collision between the clamp and tool. Appropriate clamps are required to rigidly hold the BM without collision with the tool. Adequate clamping of BM from the top as well as from the side is required to avoid vibrations so as to obtain sound welded joints.

The manufacturing and assembly of the work fixture should be easy and cost effective. The design should not only be ergonomically sound but also it should offer minimum obstruction to the tooling, machine parts, workers leading to maximum visibility and productivity. Also, the system should be robust, rigid, and should be able to serve multiple applications along with being portable. In-house robust FSW work fixture is designed and patented, which is required for firm clamping of the BM. It provides a method and a mechanism to apply clamping force not only from the top but also from the transverse and trailing sides.

Solid model of the FSW work fixture is shown in Figure 2.19.

The BMs to be welded are rigidly held with the help of specially designed clamps. The clamps are designed in such a way that they do not protrude after being mounted on the anvil so as to avoid their obstruction with the welding tool. After placing the BM on the anvil, clamps are tightened to firmly hold the BM from top.

The basic features that should be kept in mind while designing the fixture are

1. The design should be such that it accommodates both backing plate and plate to be welded.

2. Clamps should be designed in such a way that they must secure the BM from the top and at the same time not collide with the welding tool.

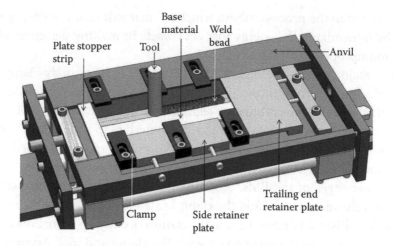

Plate stopper strip   Tool   Base material   Weld bead   Anvil

Clamp   Side retainer plate   Trailing end retainer plate

FIGURE 2.19   Solid model of FSW fixture.

3. Fixture should facilitate clamping of the BM with variable dimensions so that single work fixture can be utilized for a number of BMs with different dimensions. Also, it should be easily mounted over the bed of vertical milling machine for welding operation.

4. There should be adequate side clamping to prevent the BMs from moving away from each other during plunging of the rotating tool.

During welding, the material being stirred undergoes a complex mix of conditions of heating, SPD, and movement through action of extrusion and forging. This mix of conditions is responsible for the resulting quality of the joint. These conditions largely depend on the FSW process parameters, materials properties, tool geometry, and the thickness of material. Before combined effect of these parameters is discussed, it becomes pertinent to illustrate how these conditions get manifested in the joint in terms of various welded zone, which are ultimately responsible for mechanical properties of produced joint.

2.13 MACROSCOPIC AND MICROSCOPIC WELD ZONE IN FSW

Material processing involves combinations of strain and heat that produce desired shape and microstructure. Various phenomena such as recovery, recrystallization, and grain growth occur during material processing depending upon the working temperature and degree of deformation. Generally, material processing involves hot working and cold working of materials. Phenomena that distinguish the hot working from cold working are static recovery/static recrystallization and dynamic recovery/dynamic recrystallization. The former characterizes the cold working process and later characterizes the hot working process. FSW process involves these phenomena based on the type of material, working temperature, and degree of deformation.

2.13.1 Static Recovery

Plastic deformation results in deformation of grains due to creation of many additional dislocations, which causes increase in internal energy and change in mechanical properties of the metal being deformed. Additional dislocations are removed by heating the deformed metal to a specific temperature. Recovery process generally occurs in the range of 0.3–0.5 Tm (Tm is the melting point of metal) and restores some of the original mechanical properties with visible change in microstructure. Recovery process depends on factors such as type of metal, amount of cold work, and temperature conditions. Recovery process facilitates arrangement of dislocation within the grain and the space surrounded by the dislocation walls is known as subgrains, which differ from normal grains. The crystal orientation difference is usually greater than 10°–15° in case of grain boundaries. Therefore, grain boundaries are also known as high angle boundaries. In case of boundaries that are described by arrays of dislocations, the orientation difference between two crystals is less than 10°. These are called low angle boundaries. Recovery leads to the formation of low angle grain boundaries and development of subgrains.

2.13.2 Static Recrystallization

Deformed grains, after plastic deformation, undergo recovery process on heating at some temperature. If deformed metal is further heated to a specific temperature (around 0.5 T_m) for a certain length of time then dislocations reorganize themselves to form new, equiaxed, and unstrained grains separated from each other by high angle grain boundaries. Static recrystallization is replacement of deformed microstructure by new grains in which new grains are produced by nucleation and growth of new grains (Humphreys and Hatherly, 2002; Mishra and Ma, 2005). The temperature at which recrystallization takes place depends upon the type of metal and degree of cold work. Lower temperature is required for metal that has undergone high degree of cold work. As the amount of cold work increases, the number of dislocations and the energy associated with dislocations increase and this strain energy (ΔG_{strain}) stored in the dislocations becomes available to initiate the recrystallization (Humphreys and Hatherly, 2002). Stored strain energy (ΔG_{strain}) can be calculated by the following equation

$$\Delta G_{strain} = \alpha \rho_{dis} Gb^2 \qquad (2.1)$$

where ρ_{dis} is the dislocation density within the deformed grains, α is a constant, and b is the Burgers vector.

2.13.3 Dynamic Recovery

The dynamic recovery phenomenon occurs during deformation process and it results in the replacement of array of dislocations into subgrain structure. Strain rate (ε) and temperature (T) affect the stress (σs) during dynamic recovery. The subgrain size (d_s) developed during dynamic recovery is given by the following equation (McQueen et al., 2011):

$$d_s^{-1} = a + b \log Z \qquad (2.2)$$

where a and b are material constants.

Equation 2.2 is valid for a constant Z condition and it gives the subgrain size after achieving a steady state stress condition. Typically, pure aluminum alloys exhibit dynamic recovery whereas, for precipitation-strengthened aluminum alloys, a combination of dynamic recovery and dynamic recrystallization is observed (McQueen et al., 2011).

2.13.4 Dynamic Recrystallization

The replacement of strained grains with new grains during straining above recrystallization temperature is termed as dynamic recrystallization. Continuous straining generates new dislocations within the recrystallized grains leading to the formation of subgrains, which in turn inhibit their future growth. Further straining results in fresh nucleation sites for recrystallization. Combination of nucleation and grain growth during continuous straining results in dynamic recrystallization. Dynamic recrystallization may be continuous or discontinuous. The steady state grain size (d_s) obtained during dynamic recrystallization is expressed in terms of Z or σ (McQueen et al., 2011) in the following equations:

$$d_s^{-1} = A + B \log Z \qquad (2.3)$$

$$\sigma_s = C + D d_s^p \qquad (2.4)$$

where A, B, C, D, and p are constants.

High temperature and heavy deformation during FSW lead to recovery and recrystallization in the material. FSW process may be assumed to be a hot working technique that results in dynamic recrystallized microstructure. FS-welded joint undergoes two types of dynamic recrystallization:

1. Discontinuous dynamic recrystallization (DDRX): DDRX occur by fresh nucleation sites and grain growth of new grains.

2. Continuous dynamic recrystallization (CDRX): CDRX involves the formation of arrays of subgrains (low angle boundaries) and a gradual increase in boundary misorientations during plastic deformation leading to the development of new grains.

Heat generated in the weld softens the BM, which undergoes proper plastic deformation without reaching its melting point. The flowing material moves from AS to the RS and consolidates behind the tool pin due to forging action of the rotating tool. This plastic deformation causes change in grain structure and hence properties. Different zones in the welded joints are obtained during FSW that differ from that of fusion welding. Weld zone is classified into two categories based on microstructure and processing during FSW of aluminum alloys. These two classifications are Threadgill's (1997) and Arbegast's (2003) classification based on the microstructural zones in the welded joint and weld zone processing history, respectively.

2.13.4.1 Threadgill's Classification of Microscopic Weld Zone
The first comprehensive overview for classifying weld zones into four different microstructural zones such as SZ, TMAZ, HAZ, and unaffected BM was done by Threadgill (1997). This classification was focused on FSW of aluminum alloys and these zones are shown in Figure 2.20.

Macrostructure of welded joint shows distinct regions and interfaces between them. Typical interfaces between SZ and TMAZ at both sides of the FS-welded aluminum alloy joint are shown in Figure 2.21. Interface of SZ and TMAZ is relatively sharp, especially on the AS, where the welding

FIGURE 2.20   Microstructural zones after FSW based on Threadgill's classification.

and tool-rotating direction are similar as shown in Figure 2.21a. However, the interface of SZ and TMAZ is gradual on the RS where the welding and tool-rotating direction are opposite as shown in Figure 2.21b. Material movement around the tool pin results in the following distinct regions as shown in Figure 2.22:

- SZ or dynamically recrystallized zone (DXZ)

- TMAZ

- HAZ

- Unaffected BM

- *The SZ (also nugget, DXZ):* This zone refers to a region where stirring takes place by the tool pin. The size of SZ is slightly larger than the tool pin diameter SPD and dynamic recrystallization take place in this region resulting in grain refinement. Grains in SZ are equiaxed, and are extremely smaller in size compared to BM. Sometimes this region displays onion ring-type structure depending upon BM and process conditions. Hardness values are usually lower than the BM of heat-treatable alloys (Threadgill, 1997).

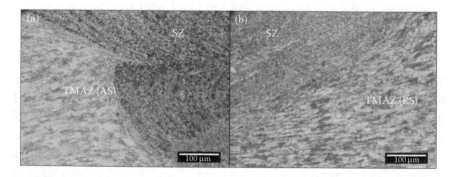

FIGURE 2.21   Typical interfaces between SZ and TMAZ of FS-welded AA7475 at AS (a) and RS (b).

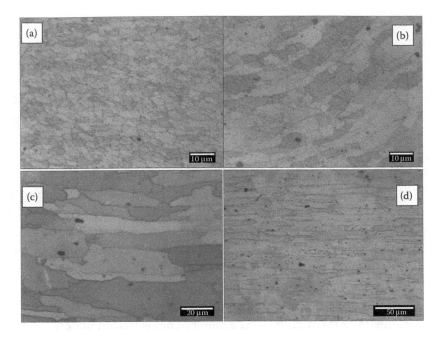

FIGURE 2.22 AA7475 FS-welded joint microstructure of (a) SZ, (b) TMAZ, (c) HAZ, and (d) base metal.

- *The TMAZ*: This zone is close to SZ at one side and HAZ at the other side. Entire deformed region other than SZ is termed as TMAZ. This region is deformed and experiences thermal cycle, which results in detectable microstructural changes. This region is affected by lower temperature and also degree of deformation compared to SZ, which indicates partial recrystallization of the grains. In this region, hardness reaches to the minimum value due to partial precipitate dissolution and grain growth caused by attainment of high temperature at some local area in TMAZ (Threadgill, 1997; Olea et al., 2006).

- *The HAZ*: It can be seen in all welding processes. This region is found in between BM and TMAZ. It does not deform but it is affected by heat and its microstructure is similar to the BM. This region generally shows reduction in the hardness profile during welding of heat-treatable aluminum alloys indicating the influence of thermal cycle.

- *Unaffected material or parent metal (BM)*: Region adjacent to the HAZ away from joint is the BM that does not deform. However, it experiences thermal cycle but not to the extent to bring significant changes in microstructure.

2.13.4.2 Arbegast's Classification for the Processing Zone

Arbegast classified the weld processing zone into five different zones such as preheat, initial deformation, extrusion, forging, and cool down zones and these are clearly shown in Figure 2.23. This classification describes the FSW as the combination of extrusion and forging processes. In the direction of welding, first is preheating zone, which experiences heat due to the thermal field around the moving tool without undergoing plastic deformation. After preheat zone, another zone is initial deformation zone, which is close to the rotating tool that is subjected to the shear stress and acts as the starting area of plastic deformation. Next is the extrusion zone in which softened material flow around the tool pin and gets extruded from leading to the trailing side of the tool. Next is the forging zone, in which moving material around the tool pin is consolidated behind the tool pin by the forging action of the tool. Last is the cool down zone, which results in the completion of the joint. The size of the zones is governed by the thermal properties of the BM and tool, process parameters and the tool geometry (Arbegast, 2003). The material being welded undergoes several processes during FSW from preheating in front of the tool to cooling behind the tool.

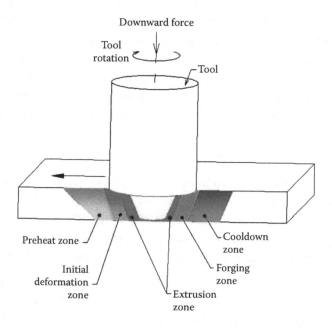

FIGURE 2.23 Arbegast's classification of the processing zones during FSW.

2.14 DEFECTS IN FSW

Improper selection of FSW process parameters results in defect formation, which in turn deteriorates the mechanical properties of the joint. The amount of heat generation affects the formation of defects in FSW. Higher heat input condition is favorable in the formation of IMCs, which are hard and brittle and reduce the ductility and strength of the joint. Higher heat input also results in grain growth and may cause the strength to suffer. At low heat input, on the other hand, insufficient softening of the material being welded occurs and flow stress increases which, in turn, leads to inefficient consolidation and inadequate mixing. Some defects such as surface galling, nugget collapsing, excess weld flash, etc. form due to high heat input and some defects such as tunneling defect, kissing bond (KB), hooking defect, joint-line remnant (JLR), etc. form due to low heat generation. Therefore, optimum heat generation plays a vital role in producing defect-free joints. Various types of defects can be found in FSW, among them some of the important defects (Figure 2.24) that severely affect the joint quality are discussed below.

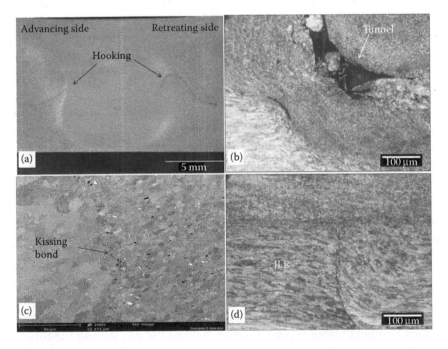

FIGURE 2.24 Defects usually observed in FS-welded joints (a) hooking defect, (b) tunneling defect, (c) KB, and (d) lack of penetration.

2.14.1 Hooking Defect

The hooking defect is one of the most probable defects that form during FSW lap joints. It is generally observed at the TMAZ toward AS when the sheets interface is pulled up into the top sheet but sometimes it may also be seen on both AS and RS. The mechanism of joining during FSW is such that the material is picked by the tool pin in the AS and during its movement from AS to RS, in the AS region, it undergoes a simultaneous upward extrusion and rotation around the pin layer by layer (Chen et al., 2008; Muthukumaran and Mukherjee, 2008). The material so dragged in the AS, when moves to the RS region undergoes downward forging action while simultaneously rotating layer by layer under the pressure of tool shoulder. This combination of extrusion and forging is mainly affected by the tool tilt and plunge depth. Improper forging of the lifted material due to inadequate tool tilt and plunge depth leads to the hooking defect as shown in Figure 2.24a.

Hook height is increased by increasing rotational speed of the tool as vertical movement of material increases (specially in case of threaded pin) and at constant rotational speed, hooking height decreases by increasing welding speed due to less vertical movement of material. Hooking defect might not appear in the HAZ as this zone does not experience plastic deformation and only gets affected by conduction heating (Cao and Jahazi, 2011).

2.14.2 Tunneling Defect

Tunneling defect is a common defect in FSW, which significantly affects mechanical properties of the welded joints. It is normally not visible on the surface as it is formed inside below the surface. Tunneling defect generally appears on the AS at the bottom of the SZ near SZ–TMAZ interface, as shown in Figure 2.24b. Like most FSW defects, tunneling also occurs due to improper selection of welding process parameters including tool rotation and traverse speed, applied pressure, tool offset, improper design of tool, etc. (Kim et al., 2006; Kumar and Kailas, 2008b; Zettler et al., 2010; Zhao et al., 2014).

It has been reported that the tunneling defects are also formed due to low heat input (Kim et al., 2006). The lower heat input results in higher flow stress causing deficient material movement around the tool pin, which hampers material consolidation behind the pin leaving the tunnel. Improper material movement and forging pressure result in tunneling defect. Insufficient forging pressure decreases maximum temperature of the weld causing flow stress of BMs to increase and consequently insufficient plastic deformation and a slugging material movement result. Under inappropriate welding conditions, the tool may move ahead before depositing enough material behind it creating a tunnel. Liu and Wu (2016) reported that insufficient material flow from RS to AS in the pin-affected zone and the shortage of downward material transfer in the shoulder-affected zone are the primary and secondary reasons for the occurrence of tunnel defect, respectively. The tunneling defect can be minimized or eliminated by appropriate generation and distribution of heat on respective sides of the joint and proper consolidation of material behind the tool pin. Tunnel acts as a site for stress concentration and also reduces the load-carrying cross-sectional area, thus, adversely affects the mechanical properties.

2.14.3 Kissing Bond

KB is a solid state bonding defect, which is formed as a result of inefficient plastic deformation of the material due to insufficient stirring of the material. If such conditions prevail the oxide layer present on the BM, if not properly removed before welding, will lead to improper bonding (Oosterkamp et al., 2004), manifesting itself as KB. This discontinuity is usually formed at the interface of the joint.

In KB, there is little or no metallic bond present between the two previously separated regions of the BM, as shown in Figure 2.24c. During the formation of the friction stir weld, it is vital that the oxide interface (specifically for aluminum alloys as it has high affinity to oxygen) between the abutting plates must be carefully removed in order to enable the formation of an intimate bond. Apart from this, lack of deposition of material, which may be associated with inadequate heat, higher flow stresses may also lead to the formation of KB (Oosterkamp et al., 2004). The lack of intimate mixing of materials at faying surface is also manifested in the form of KB. The presence of KB hinders the strong bonding between the faying surfaces of BMs, thereby resulting in poor mechanical properties. To eliminate this defect, thorough mixing during welding is required by providing material movement around the pin that should break and distribute the oxide layer on the abutting surface of the BM.

2.14.4 Incomplete Root Penetration

Inadequate tool plunge or a deficient pin length or both may cause a small portion of the joint root to remain unstirred or unconsolidated, which is manifested as a prominent incomplete root penetration. Incomplete root penetration in welded joint may also form due to local variations in the plate thickness, uneven face of abutting plates, and inappropriate tool design.

Also, incomplete root penetration occurs when the FSW pin is far away from the support anvil. Thus, an undeformed region exists between the bottom of the tool pin and the bottom surface of the plate, as shown in Figure 2.24d. When subjected to a load, the friction stir welds fail along the lack of penetration line. To eliminate this defect, thorough mixing during welding throughout the weld root is required by providing appropriate tool pin length and plunge depth so that tool pin reaches the bottom of the BM, without touching the backing anvil.

Table 2.4 summarizes the defects that are generally formed in FSW of aluminum alloys. The summary includes the location where defects are formed and the causes of their formation. Defects need to be clearly addressed and recommendations for their minimization or elimination should be made. Although FSW is a clean welding process for joining difficult-to-weld materials, the formation of defects may hamper the effectiveness of this process. With the knowledge domain on various defects

TABLE 2.4   Defects along with Their Location and Causes during FSW

S. No.	Defects	Location	Causes
1	Tunneling	At AS between the TMAZ and the SZ under the surface of the weld	1. Low plunge depth 2. Too high welding speed 3. Inappropriate pin offset 4. Too low rotational speed
2	KB	At interface in SZ	1. Improper removal of oxide layer from faying surface 2. High welding speed 3. Inadequate material movement
3	Void	At AS of the weld, on or beneath the weld surface	1. Too high welding speed 2. Improper forging pressure
4	JLR	At the root of the weld or in the SZ in the remnant of original faying surface	1. Inadequate removal of oxide later from pate edges 2. Poor tool-to-joint alignment
5	Incomplete root penetration	Below the SZ at the interface of the faying surface	1. Local variations in the plate thickness 2. Improper tool design 3. Inappropriate plunge depth
6	Hooking	TMAZ of advancing and RS in lap welding	1. Low welding speed 2. High rotational speed 3. Improper tool design 4. Inadequate tool tilt

being expanded, the FSW process may emerge to be a stronger and more reliable clean fabrication process.

2.15 MEASUREMENT OF RESPONSES FOR DEFINING WELD QUALITY

FSW is widely used in various industrial sectors owing to better joint performance and improved mechanical properties. In order to examine the quality of welded joint, various microstructural and mechanical characterization studies are required. These characterizations are needed to comprehend the in-service performance, failure behavior of the joint, and relate the effect of process parameters on the performance of the welded joint. More so, dissimilar aluminum welding by FSW exhibits heterogeneous material mixing that result in complex mechanism and thereby need to be understood.

Microstructural characterization includes the size, shape, morphology, and orientation of grains and precipitates. It also involves the fractography of fractured specimen after mechanical testing. Mechanical characterization includes the measurement of microhardness of the joint, tensile strength, elongation, impact strength, fatigue strength, etc.

2.15.1 Tensile Testing

Tensile test is performed to determine properties such as strength and ductility (through percent elongation) of a material, which set specifications to design components from the material. Continuously increasing uniaxial static tensile force is applied on specimen to determine its strength and elongation. Mechanical strengths such as proof stress, yield strength, ultimate strength, etc. can be estimated from such tests. Ultimate tensile strength (UTS), for example, is obtained by dividing the maximum applied load by the original cross-sectional area of the specimen.

$$\text{Stress (s)} = \frac{\text{Load (P}_{max})}{\text{Area (A)}} \tag{2.5}$$

Yield strength is the stress that corresponds to the yielding of the material. Tensile test can determine UTS, yield strength, and measures of ductility such as percent elongation and reduction in area. Typical stress–strain curve for aluminum 5754 is shown in Figure 2.25. It is an important test for measuring quality of welded joint. Fracture location of the tensile test sample clarifies the cause of failure that can be confirmed through

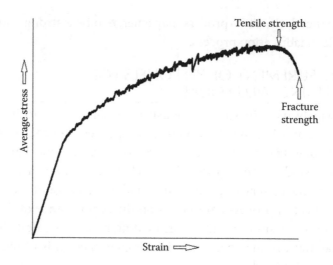

FIGURE 2.25 Engineering stress–strain curve for a typical aluminum alloy.

microstructural examination. Tensile test requires standard designed specimen as per ASTM standard. Generally, ASTM E8M is used for testing metallic samples. Tensile samples are usually prepared using wire electrical discharge machining due to its good accuracy and ability to machine any metallic material. Welded samples are tested on tensile testing machine (e.g., a tensometer). Tensometer is a device used to evaluate tensile properties (strengths and ductility) of materials. In FSW, tensile strength of joint is measured and compared with that of BM to determine the efficiency of the joint. Typical tensile specimen for metal is shown in Figure 2.26.

Different types of fractures occur depending upon the type of material, temperature, rate of loading, and state of stress. Broadly speaking, types of fracture for metal can be characterized into two categories, that is, ductile fracture and brittle fracture. Very less deformation is observed in brittle fracture and the fractured surface can also be analyzed using micrographic techniques such as scanning electron microscopy. Ductile fracture involves considerable degree of deformation, even at microconstituent

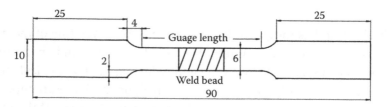

FIGURE 2.26 FSW tensile specimen.

levels, and is also manifested at the bulk of material before fracture. Brittle fractures yield little plastic deformation before fracture and such kind of failure is usually initiated from microscopic stress raisers present in the material, which can be precipitate–matrix interface, void, porosity, micro-crack of secondary phase. The boundary that distinguishes ductile and brittle fracture is arbitrary and depends upon various conditions. In poly-crystalline samples, fracture may be either transgranular (the crack prop-agates through the grains) or intergranular (the crack propagates along the grain boundaries).

2.15.2 Impact Testing

Impact test is performed under impact load to estimate the energy absorbed by the welded joint during fracture. The absorbed energy gives the tough-ness to the joint. The toughness of a material is its ability to absorb energy in elastic–plastic range until the material finally fails. It may also be esti-mated by total area under the stress–strain curve. As the test is performed under impact loading the time required to perform the test is very less and also it is very easy and economical.

Impact testing is performed to measure the ability of weld to absorb energy on shock loads as well. A standard weld transverse sample with notch is used to determine the impact strength. A hinged arm with a heavy hammer at its free end is raised at a specific height (constant poten-tial energy). Impact test sample is mounted on standard holder and after the release of arm the hammer hits the sample, breaks it, and the arm rise up with remnant energy. From the energy absorbed by the sample, its impact strength is determined. The test is performed on notched bar samples prepared in accordance with ASTM E23 standard, which speci-fies two methods namely Charpy method and Izod method. In the Charpy method, the samples are held as simple beam and in Izod method they are held as cantilever. To perform the test, welded samples are machined from the welded plates by wire electrical discharge machine (WEDM) as per stan-dard dimensions in accordance with ASTM E23 standard (typically 55 mm length and 10 mm width with a V-notch of 45° in the middle of the weld).

Charpy Impact Testing Machine is used to perform impact test. Specimen for notch impact test is shown in Figure 2.27.

2.15.3 Fatigue Testing

Fatigue refers to catastrophic failure under the action of a varying load, the magnitude of which is very less as compared to yield strength of

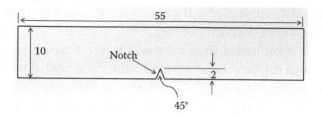

FIGURE 2.27 Specimen for notch impact test.

material. Failure under fatigue assumes importance as the failure is sudden; the magnitude at the time of failure is well below the limit and the static design for the load at fracture of the component safe. The phenomenon of fatigue failure begins with minute existing damage, which may be present either inside or on the surface of material. The damage can be a microcrack on surface or inside or it can be any other flaw. The existing damage acts as a stress raiser, which under the action of repetition of acting loads propagates into a macrocrack without setting any alarm of danger. Once the size of the propagated crack becomes so large that the remaining cross-sectional area becomes insufficient to support the prevailing load the component fails suddenly. Thus, testing a welded part for its failure under fatigue becomes most important. Failure under fatigue occurs in three stages: (i) crack initiation, (ii) propagation of crack, and (iii) sudden abrupt failure.

Testing for fatigue requires the standard test specimen to be highly polished as the surface roughness marks may act as flaws behaving as stress concentrators. The test may be performed under different types of loading and one of the common conditions of load is the "constant amplitude alternating axial load." The specimen preparation and test procedure under this type of load is prescribed in ASTM E466 standard. This test is performed on axial fatigue testing machining. Other types of fatigue tests include bending fatigue test, tensional fatigue test, etc. Bending fatigue tests are performed on high rpm rotating specimen subjected to bending loads. The specimen for the bending fatigue tests are of circular section.

Irrespective of the test method, for testing fatigue strength (known as endurance limit), a series of tests are performed on a number of standard fatigue test specimens at different variable stress σ levels. The stress endured is then plotted against the number of cycles N sustained by the specimen. The σ-N data (the stress and the number of cycles to failure) are plotted in a curve, which are generally referred to as the σ-N curve of S-N curve of material. The S-N curve is indicative of the fact that smaller numbers of cycles of load would be needed to cause failure at high stress levels and vice versa. This curve also establishes fatigue limit (endurance limit) as a theoretical stress for which the specimen would never fail regardless of the number of applied cycles of load. Generally, the fatigue limit is established from S-N plots as a stress level for which the specimens would not fail for 1–10 million cycles.

2.15.4 Preparation for Microstructural Investigation

Friction stir welded sample needs to undergo microstructural examination to know the location of internal imperfections and defects, analyze reasons of low strength and understand the mechanism of material flow. For analysis of microconstituent, grain structures and morphology, optical microscopy is important. However, advanced examination and research often require scanning electron microscopes (SEM), electron dispersive spectroscopy, x-ray diffractometers, and possibly other scanning devices as well.

For all these types of microstructural analyses, special procedures for sample preparation are adopted. Standard procedure requires that samples for microstructural examination need to be properly polished and sometimes etching is also required based on the type of requirement. Sample preparation is performed through a sequence of operations on the sectioned samples involving grinding and polishing followed by etching with some suitable reagent. The samples thus prepared are ready for examination through optical microscope and other suitable devices.

Steps in sample preparation for microstructure study.

1. *Sectioning*: Various methods can be used for making microstructure samples including hand sawing, milling, laser cutting, water jet cutting, wire electric discharge machining, etc. These operations alter the surface properties and microstructure of the sample. Conventional machining produces severe cold work, which can alter the microstructure of a sample and laser or jet cutting generate heat that may alter the properties of welded joint. Wire electric discharge machining (sectioning) offers the best solution to minimize these undesirable features; the resultant surface is smooth, and the sectioning task is quickly accomplished (although WEDM also alters the surface properties of the sample, which can be removed using belt grinder before polishing).

2. *Mounting*: Samples for microstructural examination are usually hot mounted in Bakelite or other cold-mounting compound for easy handling, and maintain uniformity during grinding and polishing. Molding of specimen can be done either by hot mounting press or by using cold mount kit manually. Hot mounting press involves temperature and pressure that may affect the microstructure of the welded joint. Hence, cold mounting is preferred for materials with low melting point.

3. *Grinding*: This stage involves coarse, medium, and fine grinding of the prepared samples. Initially, samples are grinded on belt grinder to remove the effect of WEDM from the weld surface and then fine grinding is done on metallurgical grinding and polishing machine. Each stage of metallographic sample preparation must be carefully performed; the entire process is designed to produce a scratch-free surface by employing a series of successively finer abrasives. The idea is to carefully move from one stage to the next where the abrasives become finer at each successive stage. Movement from one stage to the next should only proceed when all of the scratches from the preceding stage are completely removed. The pieces are wet grinded using 100, 200, 400, 600,800, 1000, 1200, 1500, and 2000 grit silicon carbide paper. Care must be taken to ensure that the grinding in stage is performed with a constant force and it should be done for a constant duration of time.

4. *Polishing*: Polishing is done after grinding by using SiC abrasives. Polishing involves the use of 0.5 μm diamond paste or 0.05 μm size colloidal silica to get mirror polished surface of the weld for better microstructural examination.

5. *Etching*: Microscopic examination of a properly polished, unetched specimen will reveal only a few structural features such as inclusions and cracks or other physical imperfections. Etching is used to highlight, and sometimes identify, microstructural features or phases present. Even in a carefully prepared sample, a surface layer of disturbed metal, resulting from the final polishing stage, is always present and must be removed. Etchants are usually dilute acid or dilute alkalis in water, alcohol, or some other solvent. Etching occurs when the acid or base is placed on the specimen surface because of the difference in rate of attack of the various phases present and their orientation.

2.15.5 Microhardness Measurement

Hardness is generally defined as resistance to indentation and sometimes resistance to deformation. Particularly for metals, hardness is the measure of their resistance to plastic deformation. Different hardness measurement tests are available but indentation hardness test is mostly used for metal in engineering application.

In Vickers hardness test, square-base diamond pyramid is used as indenter with included angle between opposite faces of 136°. It is usually called as diamond pyramid hardness (DPH) test owing to the shape of indenter. It is defined as the load applied by the indenter divided by the surface area of indentation. It may be determined by the following equation:

$$DPH = \frac{2P\sin(\theta/2)}{L^2} = \frac{1.854P}{L^2} \qquad (2.6)$$

where
  P = load applied (kg)
  L = average length of diagonals (mm)
  θ = include angle between opposite faces of diamond (136°)

It is widely used in research work and provides hardness values for large range of materials from very soft metals with DPH of 5 to hard metals with DPH of 1500 with the same indenter. It is not preferred for routine testing because it is a slow process and sample preparation is required.

For correlation between microstructure and mechanical properties of welded joints, hardness over very small area is required. Measurement of hardness for individual constituent of microstructure is termed as microhardness. Microhardness testing is a method for measuring the hardness of a material on microscopic scale. A precise diamond indenter is impressed into the material at loads from 10 g to 200 g for 20 s during testing. The impression length is measured microscopically, and the test load is used to calculate a hardness value. The term microhardness test usually refers to static indentations made with loads not exceeding 1 kgf. The test samples should have a smooth surface and be held perpendicular to the indenter. The specimens for microhardness measurements are sectioned by wire electro discharge machining to the required size and then polishing is performed using different grades of emery papers

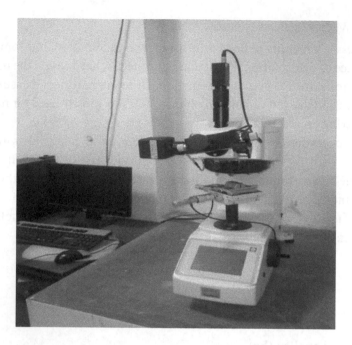

FIGURE 2.28   Mitutoyo microhardness testing machine.

on metallurgical polishing machine. The test load used for microhardness measurement depends on the resolution and hardness of the material. Typical machine set up for microhardness measurement is shown in Figure 2.28.

CHAPTER 3

Friction Stir Welding of Aluminum Alloys

3.1 OVERVIEW

FSW is a solid state welding process, which was originally developed to weld aluminum alloys. However, over the years, it evolved as an excellent fabrication process for other materials such as magnesium, copper, titanium, steel, etc. Consumables such as filler materials and shielding environment are not used in this process. A unique feature of this process is that it produces joints without internal imperfections such as porosity. Sound aluminum joints in thickness ranging from less than 1 mm to over 60 mm can be made in single pass. Several properties of joints fabricated using FSW are better than joints made by fusion welding. The potential of FSW was tapped by industries that have large intrinsic use of aluminum alloys such as aerospace, shipbuilding, and automotive industries. Although it was invented in 1991, its large-scale commercialization commenced beyond 1997 when the aircraft and aerospace major, Boeing, exploited the potential of this process and adopted it to fabricate rocket fuel tanks. Significant gains in terms of both productivity and cost (cost with FSW was mere 20% as compared to cost with riveting) were achieved, which led to the application of FSW for fabrication of other components such as carrier beams, fuselages, and wings.

The enormous potential of FSW attracted other industrial sectors and, shipbuilding industry became a major sector in which weight of aluminum being FSWed is one of the largest. Apart from joint properties, other biggest advantages of FSW are its high productivity, virtually no postweld finishing, distortion correction, etc. Development of high-strength aluminum alloys has eliminated the notion of "Aluminum being not as strong as steel." Currently, aluminum alloys are available that are stronger than steel (e.g., AA2214). Several high-strength aluminum alloys have never been used in shipbuilding, because of their poor weldability, but this barrier was successfully overcome by FSW. Another high-volume welding activity sector, that is, the automotive sector, also employed FSW and its use is spreading in geometric proportions. Fast-changing and ever-increasing customer demands are to be fulfilled by this sector for which parts made of different materials were required to be fabricated and to accomplish this, FSW is proving to be a perfect choice. Several heat-treatable and non-heat-treatable alloys and their combinations can be joined by FSW with ease and at a speed of several meters per minute. Typically, AA6082 is a difficult-to-weld alloy by fusion process, but its similar and dissimilar joining with several other alloys (e.g., with 2xxx, 5xxx, and 7xxx series) can easily be performed by FSW. All these characteristics of FSW have made it an attractive choice for automotive sector and it finds use in fabricating more than 15 critical components including cylinder heads, axles, intake manifold, etc. and the use continues to increase further.

3.2 PROBLEMS RELATED TO THE WELDING OF ALUMINUM ALLOYS

Several aluminum alloys can be generally welded using fusion welding processes such as GMAW and gas tungsten arc welding (GTAW). But, defects, mechanical and metallurgical discontinuities, and quality of joint often remain a major concern for some alloys that are heat treatable and/or their intended use is in corrosive environment.

A major problem during aluminum welding is *cracking* at the weld, which is mainly due to low strength and ductility at elevated temperature in the weld zone and in the heat-affected zone. This behavior is commonly called hot shortness. Hot cracking is a very sensitive issue in fusion welding of aluminum alloys, which largely depends on the filler material composition, joint area cleanliness, shielding environment, etc. Also, simultaneous presence of magnesium and copper in an aluminum weld is detrimental as it affects hot cracking. All fusion welding processes use an external filler metal and its composition not only affects hot cracking but also the corrosion resistance of weld. For specific corrosive environments, a particular filler metal composition matching with base metal chemistry is used. There is strict recommendation for either using or not using filler metals during fusion welding of aluminum alloys. It has been observed that even the best available fusion welding process for joining aluminum alloys is not capable of joining all aluminum alloys.

Alloys that are difficult to join include most 2xxx and several 7xxx alloys. Users have found solution to welding problems of such alloys in solid state welding such as FW, diffusion bonding, explosive welding, FSW, etc. These solid state welding processes are also suitable for welding aluminum alloys to other metals and alloys that are metallurgically not compatible. The typical joint quality requirements for applications where these alloys are used most, necessitated rapid developments in process suitable for this purpose, and among those FSW remains a distinguished process.

All aluminum alloys suffer from a major problem of solidification cracking in weldment with some risk of liquation cracking in HAZ region. Great efforts are needed to check solidification cracking by way of avoiding weldment composition whose solidification range is close to solid solubility limit of primary alloy composition. The problem of solidification cracking is more severe in 2xxx, 7xxx, and 8xxx alloys. A problem associated with liquid-phase dissolution of hydrogen causes hydrogen embrittlement in weldments, which severely reduces weld ductility. There is vast difference in solubility of hydrogen in liquid and solid states (nearly 50:1) and the drop of solubility with drop in temperature is very steep near solidification temperature; this creates minute porosity at grain boundaries making the weld brittle. To avoid embrittlement, hydrogen in all forms including hydrocarbons and moisture should be strictly avoided from being near to the weld.

FSW being solid state in nature provides excellent remedy to most welding problems of aluminum, yet it is not free from challenges as has its own typical defects if the welding is not performed effectively. Weld consolidation in FSW is obtained through stirring in which SPD is the predominant mechanism. If the stirring is not performed at adequate heat, it causes serious problems related to bonding and closeness of contiguity at interatomic, intermolecular, and intergranular levels. This is what is termed as inadequate bonding, which is very difficult to identify through any nondestructive test including radiography, but it is severely detrimental to mechanical properties. Stirring action and a resultant material flow largely depends on the flow stress of the material being deformed, which in turn depends on the prevailing temperature. In case of insufficient heat, the material flow may remain deficient and may result in what is known as tunneling defect. Tunneling defect occurs in the subsurface region but can be detected through nondestructive tests. Tunneling defect also adversely affects the mechanical properties of the joint.

Lack of heat and the presence of oxide layer at the joint face prior to welding causes another serious problem that is manifested in the form of a defect known as KB. KB is characterized by a meek bonding of material at the AS–RS interface. The KB also lowers the mechanical properties significantly and cannot be detected even by nondestructive tests such as radiography. It is required that the joint face must be perfectly cleaned just before welding and the welding should be performed at adequate heat. Most of these FSW joining issues can be tackled by choosing correct combinations of parameters, most important of which are tool rotational speed, welding speed, tool profile, tilt angle, and plunge depth. Too low or too high values of most of these parameters may have adverse effect. For example, in case the plunge depth is too low, it may cause lack of penetration, which may result in a common defect called JLR. The JLR defect lowers the mechanical properties and it can be detected through nondestructive tests. Extremely high values of plunge depth may result in excessive flash formation and thinning at the weld bead. Likewise, inappropriate parameter combination that causes overheat may result in dissolution of precipitates and grain coarsening and both resulting in lowering of strength. The problems arising from inadequate parameter combination makes the situation even more complex when dissimilar materials are welded together.

3.3 FSW OF ALUMINUM ALLOYS

FSW is widely used to join aluminum alloys in many industrial sectors including automobile, shipbuilding, aerospace, etc. Aluminum alloys possess several properties that make them best alternatives among the available materials for transportation and commercial applications. Unfortunately, several aluminum alloys show poor weldability by conventional fusion welding processes. Use of aluminum alloys is limited due to the problems related to their weldability. Development of FSW addressed most of the problems related to their welding and due to this, application of these alloys in several industrial sectors is gaining momentum. FSW of aluminum alloys resulted in fruitful use of properties possessed by aluminum alloys and also good joint strength is achieved. At present, aluminum alloys are widely welded by using FSW by various industries worldwide.

Although FSW possesses several merits, its application in joining aluminum alloys needs special attention in terms of understanding the process and also the effect of various FSW process parameters, including tool geometry, on the quality of welded joint. This is important in view of the vast difference in mechanical properties of aluminum alloys available for use as the response of each alloy to FSW process is often different. Based on major alloying element and chemical composition, aluminum alloys are classified into eight different series. Broadly, wrought aluminum alloys are classified into two groups: (i) heat treatable and (ii) non-heat treatable. Many researchers have performed FSW of different series of aluminum alloys.

1xxx, 3xxx, and 5xxx series aluminum alloys belong to non-heat-treatable category and their strength is not affected by the second-phase particles. The strength of these alloys is generally enhanced by strain hardening. Strain hardened aluminum alloys are designated by "Hxxx," where xxx indicates the amount of cold working (Mishra and Ma, 2005). 2xxx, 6xxx, and 7xxx series aluminum alloys come under the ambit of heat-treatable alloys, which are strengthened by second-phase particles as well. The presence of precipitates in these alloys enhances their strength and therefore, they are also known as precipitation hardenable aluminum alloys.

Due to the difference in strengthening mechanism, the effect of FSW on the joint quality of both categories of aluminum alloys is different. Heat input and plastic deformation associated with FSW gives different microstructural features and mechanical properties of various weld zones of heat-treatable and non-heat-treatable aluminum alloys. Generally, the welding efficiency of non-heat-treatable alloys is greater than that of heat-treatable alloys. FSW results in some loss of prior cold work in SZ, TMAZ, and HAZ, grain refinement in SZ; and coarsening in TMAZ and HAZ during welding of non-heat-treatable aluminum alloys. On the other hand, dissolution of strengthening precipitates and grain refinement in SZ, coarsening of precipitates and grains in TMAZ and HAZ, are usually observed in heat-treatable alloys.

Among all aluminum alloys, AA2xxx, 5xxx, 6xxx, and 7xxx series are widely used in automobile, aerospace, shipbuilding, and other structural industries. FSW of aluminum alloys of these series, therefore, need special attention. Probably, this is the reason that large number of researchers have investigated welding of 2xxx, 5xxx, 6xxx, and 7xxx aluminum alloy series. In the following section, FSW of different aluminum alloys has been summarized briefly.

3.4 FSW OF 2XXX SERIES ALUMINUM ALLOYS

AA2xxx series are heat-treatable (precipitation hardenable) alloys with copper as a major alloying element. Precipitation sequence for Al–Cu alloys is given as

$$\alpha \,(\text{Supersaturated}) \rightarrow \text{G.P. zone} \rightarrow \theta'' \rightarrow \theta' \rightarrow \text{stable } \theta$$

where α is the Al matrix phase while θ'' and θ' are the metastable phases and θ is stable phase.

Many researchers investigated microstructure and mechanical properties of friction stir welded 2xxx alloys and correlated the evolved microstructure with mechanical properties. FSW of some 2xxx aluminum alloys is presented in Table 3.1. Mechanical properties and microhardness depend on the features of evolved microstructure of weld such as grain size, grain morphology, dislocation density, distribution, size of strengthening precipitates, etc. FSW of 2xxx alloys in T87 condition results in dissolution of θ'' precipitates and coarsening of θ' and θ precipitates in SZ, dissolution of θ' precipitates in TMAZ, and coarsening of θ' and θ precipitates in HAZ.

FSW of AA2219 was performed to investigate the deformation mechanism and texture evolution of different weld zones (Chen and Jiang, 2014). Different microstructural features were obtained for different regions: elongated grains were observed in base metal, HAZ, and TMAZ, whereas SZ was characterized by fine equiaxed grains. Dynamic recovery occurred in TMAZ, whereas SZ underwent geometric dynamic recrystallization (GDRX). Fine grains were formed in SZ by strain-induced boundary migration mechanism (Chen and Jiang, 2014). During FSW of Al alloy, CDRX and DDRX occur in SZ due to high stacking fault energy of aluminum. Generally, CDRX takes place in SZ of FSWed aluminum alloy and sometimes DDRX also occur (Humphreys and Hatherly, 2004).

FSW of 2219-T6 was investigated using the technique of reverse rotation of tool pin with respect to tool shoulder (Li and Liu, 2013). Welding speed affected the microstructure and mechanical properties of welded AA2219 at constant rotational speed. Effect of welding speed on the weld macrostructure can be seen in Figure 3.1. Different welding speeds (50, 100, and 150 mm/min) resulted in the formation of benign

TABLE 3.1   FSW of Some 2xxx Aluminum Alloys

Material	Thickness (mm)	Rotation Speed (rpm)	Welding Speed (mm/min)	Pin Profile	References
2219-T6	5.6	800	100, 400, 800	Frustum cone-like threaded	Zhang et al. (2015)
2219-T62	6	600, 1500	60	Threaded cylindrical	Chen et al. (2015)
2024-T3	3.175	600	200	Threaded cylindrical	Zhang et al. (2015)
2198-T851	2.5, 3.2, 4.5	600	300	Conical threaded with three facets	Tavares et al. (2013)
2017	6	653, 910, 1280	67, 86, 109	Tapered cylindrical	Bouzaiene et al. (2012)
2524-T3	2	900	300	–	Fu et al. (2012)
2519-T87	10	250	30	Tapered threaded	Liang et al. (2012)
2219-T6	5	800	50, 100, 150, 200	Conical threaded	Li and Liu (2013)
2024-T35	5	900, 1120, 1400	16, 40, 80	Triangular and square	Abd El-Hafez (2011)

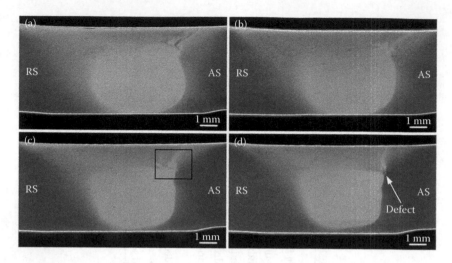

FIGURE 3.1 Macrostructure of joints welded at different welding speeds: (a) 50, (b) 100, (c) 150, and (d) 200 mm/min. (With kind permission from Springer Science+Business Media: *International Journal of Advanced Manufacturing Technology*, 68, 2013, 2071–2083, J.Q. Li, H.J. Liu.)

weld (Figure 3.1a–c). Increase in welding speed in the range from 50 to 150 mm/min gradually lowered the heat input per unit length, which led to reduction in the size of each zone. Also, increase in the welding speed leads to decrease in shoulder-affected zone as well as pin-affected zone. Defect-free joints were obtained up to welding speed of 150 mm/ min. Void-like defect was observed at the AS of weld when the welding speed was further increased to 200 mm/min, as shown in Figure 3.1d, mainly due to insufficient heat input and a resulting inadequate defor-mation of material around the tool pin. Size of grains in SZ depends on degree of plastic deformation and amount of heat input. High welding speed resulted in relatively fine and equiaxed grains in SZ due to less holding time to grow the recrystallized grains. Effect of welding speed on microhardness distribution across the weld section is shown in Figure 3.2, which reveals asymmetric distribution of microhardness on either side of welded joint due to different material flow mechanism and shear strain (Li and Liu, 2013). Increase in welding speed from 50 to 200 mm/min resulted in gradual increase in microhardness of TMAZ and HAZ on both sides of the weld due to restrictions in grain growth and precipitate dissolution caused by decrease in the holding time and peak temperature.

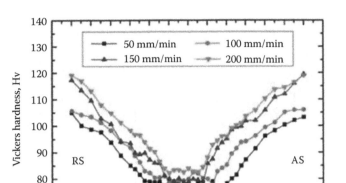

FIGURE 3.2 Microhardness profile of the welded joints at different welding speeds. (With kind permission from Springer Science+Business Media: *International Journal of Advanced Manufacturing Technology*, 68, 2013, 2071–2083, J.Q. Li, H.J. Liu.)

Tensile strength of welded joints first increased gradually with welding speed up to 150 mm/min and then decreased. Joint fabricated with 150 mm/min welding exhibited maximum tensile strength of 328 MPa (73.7% of BM) and that with 50 mm/min resulted in lowest tensile strength due to grain growth caused by higher heat input and holding time (Li and Liu, 2013).

FSW of AA2014-T351 was investigated to observe the effect of "rotational speed to welding speed (R/v) ratio" on mechanical properties of welded joints (Radisavljevic et al., 2013). Values of R/v in the range 5.0–10.27 were tried and maximum joint efficiency of FSWed joint (82.3%) was observed at R/v ratio of 10.17. Tensile strength of the joints decreased by increasing welding speed while keeping the rotational speed constant and this happened due to decrease in peak temperature and the time required for adequate material movement around the pin. R/v of 10.17 showed maximum tensile strength due to the generation of adequate heat required for sufficient plastic deformation of the material being welded. Tensile specimens fractured at the interface of SZ and TMAZ of the welded joint. Joint efficiency at different R/v values is shown in Figure 3.3. Joint fabricated using R/v ratio of 10.17 and 10.27 exhibited the maximum average tensile strength of 82.3 and 81.9%, respectively. Poor tensile strength of the joint

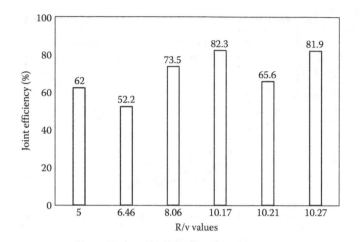

FIGURE 3.3 Joint efficiencies of welded joints at different R/v values. (From I. Radisavljevic et al. *Transactions of Nonferrous Metals Society of China*. 23:3525–3539, 2013.)

fabricated with R/v value of 10.21 (close to 10.17 and 10.27) may be due to inappropriate relation between heat input and flow of material (Liu et al., 2005; Di et al., 2006).

The width and depth of the SZ were observed to be affected by R/v ratio and they increased with the increase in the value of the ratio. Larger R/v ratio resulted in higher heat input and stirring increment, which improved the flow of material around the pin. This led to the generation of larger frictional heat, which helped in easy flow of material and resulted in larger SZ (Hassan et al., 2003; Zhang et al., 2011).

FSW of AA2519-T87 was investigated in which precipitate distribution and microhardness of welded joints were studied (Liang et al., 2012). The researchers also discussed the distribution and size of strengthening precipitates in various zones of welded joint. W-shaped microhardness profile of welded joint was plotted, which showed that the microhardness in all zones was lower as compared to BM. Lowest microhardness was found in the middle of TMAZ on either side of the joint. Microhardness decreased steeply toward HAZ from BM and then it again decreased upto TMAZ. By further increasing the distance from TMAZ to weld center, microhardness was found to increase drastically. They attributed variations in microhardness in SZ, TMAZ, and HAZ zones to variation in grain size. Lowest microhardness in TMAZ was attributed to dissolution of strengthening precipitates and coarsening of recrystalized grains.

Optical micrograph of FSWed AA2519-T87 is shown in Figure 3.4. Grain refinement was observed in SZ due to dynamic recrystallization, as shown in Figure 3.2a, and size of these fine grains increased from SZ to TMAZ, as shown in Figure 3.4a–c. TMAZ experienced heat and deformation during welding and partial recrystallization occurred in TMAZ due to insufficient deformation strain. Therefore, elongated and deformed grains were observed in TMAZ, as shown in Figure 3.4d. HAZ experienced only

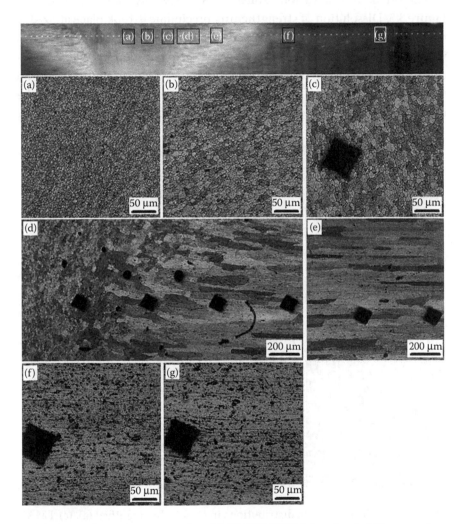

FIGURE 3.4 Optical micrographs of 2519-T87 FSW joints (a–c) different regions of SZ, (d) SZ–TMAZ interface, (e) TMAZ, (f) HAZ, and (g) BM. (Reprinted from *Materials and Design*, 35, X. Liang et al. Study on the microstructure in a friction stir welded 2519-T87 Al alloy. 603–608, Copyright 2012, with permission from Elsevier.)

thermal cycle and it was characterized by the grains morphology similar to the BM (Figure 3.4f and g).

The precipitation sequence was established in an Al–Cu alloy as

Supersaturated solid solution → Quenched clusters → G.P. (I) → G.P. (II) → θ'' → θ' → Stable θ (Son et al., 2005).

Microstructure of various zones of AA2519 FSW joint is shown in Figure 3.5. Dissolution of strengthening precipitate θ' took place at SZ due to shear and high heat input and coarse θ precipitates with around 400 nm in diameter were observed (Figure 3.5a). Precipitates looped by the dislocations, as shown in Figure 3.5b, which obstruct the motion of dislocation resulted in the improvement of material strength. Figure 3.5c–e shows the microstructures at different location of TMAZ, that is, close to SZ, middle,

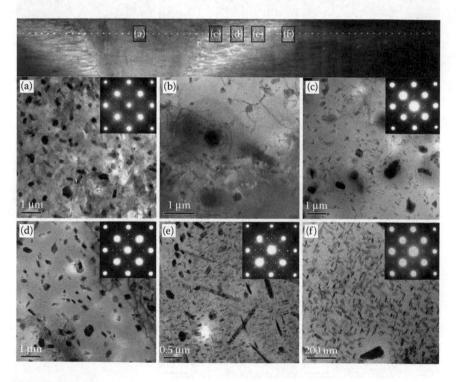

FIGURE 3.5 The precipitate distributions in (a) SZ, (b) magnified (a), (c) TMAZ close to SZ, (d) middle of TMAZ, (e) TMAZ close to HAZ, and (f) HAZ. (Reprinted from *Materials and Design*, 35, X. Liang et al. Study on the microstructure in a friction stir welded 2519-T87 Al alloy. 603–608, Copyright 2012, with permission from Elsevier.)

and close to HAZ. Dislocation density and precipitate distribution is different in TMAZ from SZ, as shown in Figure 3.5c, and precipitates are significantly larger compared to BM. Dislocations and θ' precipitates are absent in the middle of TMAZ and only coarse stable θ precipitates are observed (Figure 3.5d). θ' and θ precipitates are observed in microstructure of TMAZ close to HAZ, as shown in Figure 3.5e. Density of θ' precipitates in TMAZ close to HAZ is found to be higher than that in TMAZ close to SZ, which resulted in gradual increase in microhardness from the middle of TMAZ to the HAZ. Homogeneously distributed larger θ' precipitates (80 nm diameter, 20 nm thick) with lower density are observed in microstructure of HAZ, as shown in Figure 3.5f. The size of θ' precipitates decreases from HAZ to BM leading to higher microhardness values while going toward BM from HAZ.

3.5 FSW OF 5XXX SERIES ALUMINUM ALLOYS

FSW of 5xxx aluminum alloys are widely used in shipbuilding, automobile, etc. This series of alloys have Mg as a major alloying element, they are non-heat-treatable (strengthened by strain hardening) and possess good corrosion resistance among non-heat-treatable aluminum alloys. The presence of magnesium in 5xxx alloys enhances resistance to especially seawater corrosion. FSW of 5xxx results in the refinement of grains at SZ and coarsening of grains in TMAZ and HAZ, which causes higher microhardness in SZ compared to TMAZ and HAZ. Mechanical properties of

TABLE 3.2   FSW of Some 5xxx Series Aluminum Alloys

Material	Thickness (mm)	Rotation Speed (rpm)	Welding Speed (mm/min)	Pin Profile	References
5083	3	1800	1000	Tapered cylindrical	Rao et al. (2013)
5A06	108	610	60	Conical	Guo et al. (2017)
5083-H116	5	200	300	Threaded cylindrical	Attallah et al. (2007)
5754-O	2.3	500	500	MX triflute	Attallah et al. (2007)
5083-H18	1.64, 1.24	1000, 1500	100–450	Threaded cylindrical	Gan et al. (2008)
5083-O	9.5	–	76.2	Threaded cylindrical	Long et al. (2007)
AA5083	3	–	100, 150, 200	Threaded cylindrical	Peel et al. (2003)

these alloys heavily depend on dislocation density and grain size. FSW of some 5xxx series aluminum alloys is presented in Table 3.2.

During FSW of 5A06 alloy, researchers have observed five distinct zones based on the difference in microstructure viz. shoulder-affected zone (SAZ), SZ, TMAZ, HAZ, and BM (Guo et al., 2017). SAZ is the uppermost area of nugget, which is stirred by rotation of tool shoulder. Microstructure of HAZ is similar to that of BM but coarsening of grains takes place due to thermal cycle during welding. TMAZ undergoes intense plastic deformation with partial recrystallization, which is evident as a clear boundary between SZ and TMAZ. Nonuniform distribution of grains was observed in SAZ, whereas fine equiaxed grains with onion ring structure were observed in SZ. Grain size in SAZ is generally larger than that in SZ due to larger rubbing area of shoulder with the material being welded, which leads to higher frictional heat generation and longer heat-holding time resulting in coarsening of grains and reduction in dislocation density. Dislocation density is lower in SAZ compared to that of BM, as shown in Figure 3.6. The presence of dislocation pile up and subboundary restrict further movement of dislocation, which improves the mechanical properties of the material. Lower portion of weld nugget is stirred by tool pin, which creates less frictional heat and less heat-holding time resulting in smaller grains.

Transmission electron microscopy (TEM) analysis of SZ performed by the researchers reveals increased dislocation density of SZ as compared to SAZ, as shown in Figure 3.6d–f, due to the movement of dislocation caused by higher heat input in SAZ (Guo et al., 2017). Average grain size increases from the lower portion of nugget (SZ) to the upper portion of nugget (SAZ), which leads to reduction of microhardness in SAZ compared to SZ. Also, based on Hall–Petch relationship, microhardness of SZ is higher than that of TMAZ and HAZ due to increase in grain size on moving from SZ to HAZ.

FSW of AA5083 was investigated and different microstructure and mechanical properties on either side of the SZ was observed due to difference in degree of plastic deformation on AS and RS. SZ/TMAZ interface at AS is sharp, whereas interface is dull at RS (Krishnan, 2002). He observed that different onion ring structure was formed in SZ at AS and RS, in general. Clear onion ring structure was observed in SZ close to AS and not well-defined onion ring structure was observed in SZ close to RS. Abrupt change in microstructure was observed at the interface of TMAZ and SZ at AS, which was also observed by Klotzer et al. (2006) but TMAZ/SZ interface at RS was much more diffused. In other words, transition line

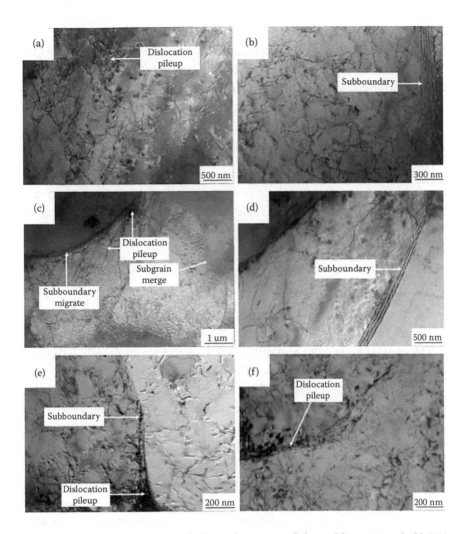

FIGURE 3.6 TEM images on different locations of the welding joint: (a, b) BM; (c, d) SAZ; and (e, f) SZ. (Reprinted from *Materials and Design*, 113, N. Guo et al. Microstructure and mechanical properties in friction stir welded 5A06 aluminum alloy thick plate. 273–283, Copyright 2017, with permission from Elsevier.)

between SZ and TMAZ at AS is clear and it is difficult to define transition line at RS. Abrupt change in microstructure at AS results in rapid change in microhardness from HAZ to SZ at AS while gradual change is observed at RS. Asymmetric hardness profile was observed on either side of SZ (Krishnan, 2002).

Krishnan (2002) further observed that different microstructure at AS and RS of SZ caused variation in tensile strength and elongation on either side of SZ. He measured tensile strength and elongation at different zones on both sides of SZ by using microtensile testing. Slicing and testing of microtensile specimen at different location from SZ at AS and RS showed different values of strength and ductility. Ductility at AS increased more rapidly compared with that at RS as AS experienced larger temperature compared to RS.

3.6 FSW OF 6XXX SERIES ALUMINUM ALLOYS

6xxx series aluminum alloys are heat-treatable Al–Mg–Si alloys, which are widely used in structural applications. FSW of some 6xxx series aluminum alloys is presented in Table 3.3. Strength of 6xxx alloys can be improved by strengthening precipitates whose sequence is given below:

$$\alpha\,(\text{Supersaturated}) \rightarrow \text{G.P. zone} \rightarrow \beta'' \rightarrow \beta' \rightarrow \beta\,(\text{Stable})$$

where β'', β' are metastable phases and β (Mg_2Si) is equilibrium phase. Such sequences occur for high Mg:Si ratios (Chakraborti and Laughlin, 2004).

Mechanical properties and microstructure of FSWed AA6082-T6 using various tool shoulder geometries (fillet + scroll tool, fillet + cavity tool, and fillet tool) were investigated by Scialpi et al. (2007). They found that the type of tool shoulder affected the material movement under shoulder leading to crown-shaped appearance at FSWed joint. Tool with fillet and scroll resulted in smooth crown with some flash; smooth crown with very little flash was obtained in the joint fabricated with tool having fillet and cavity; and similar crown was obtained for tool with fillet only.

Scialpi et al. (2007) also found that SZ contained fine and equiaxed grains due to dynamic recrystallization and SZ grain size was considerably smaller as compared to BM. Different shoulder geometries showed variation in material movement under the shoulder leading to difference in SZ dimension. Strength and microhardness in FSWed 6xxx series alloys with T6 condition depends on size and distribution of strengthening precipitates (e.g., Mg_2Si) and size of grains in defect-free joints. FSW results in softening of welded region leading to reduction in microhardness. Lower hardness in SZ was also caused by dissolution of β'' (Mg_5Si_6) precipitates, which is responsible for hardening in the BM and some hardness in SZ is recovered due to grain refinement (Svensson and Karlsson, 1999).

FSW of AA6016-T4 was investigated using conical and scrolled shoulder (Rodrigues et al., 2009). TEM images of SZ of joint fabricated using conical tool and scrolled tool as depicted in Figure 3.7a and b, respectively,

TABLE 3.3   FSW of 6xxx Series Aluminum Alloys

Material	Thickness (mm)	Rotation Speed (rpm)	Welding Speed (mm/min)	Pin Profile	References
6082 T6	1.5	1810	460	Cylindrical	Scialpi et al. (2007)
6016-T4	1	1120, 1800	180, 320	Threaded cylindrical	Rodrigues et al. (2009)
6061	6	800, 1000, 1200, 1400, 1600	75	Cylindrical, tapered, threaded, square, triangular	Elangovan et al. (2008)
6082	4	1600	40–460	Cylindrical	Cavaliere et al. (2008)
6111-T4	2.67, 1.54	1000, 1500	100–600	Threaded cylindrical	Gan et al. (2008)
6056-T6	10	1800	900	Truncated cone pin	Cabibbo et al. (2007)
6056-T4, T6	4	1600	800	–	Olea et al. (2007)

showed huge difference in average grain sizes in SZ for both the joints. Average grain size of 7 μm was observed in conical tool weld while that of 2 μm in scrolled tool weld. Size of recrystallized grains depends on the amount of heat generated and degree of plastic deformation. Higher degree of plastic deformation and lower heat input results in finer grains. Joints fabricated with scrolled tool results in larger plastic deformation levels and low heat generation leading to finer grains. Also, scrolled tool weld showed high density of coarse precipitates than conical tool weld, as is evident from Figure 3.7a and b. Lesser precipitates in conical weld may be due to higher peak temperature that causes dissolution of precipitates. Many researchers proposed similar mechanism of softening of HAZ and TMAZ due to dissolution and coarsening of precipitates in 6xxx series alloys (Sato et al., 1999a,b; Heinz and Skrotzki, 2002).

FSW of AA6061 was investigated at various rotation speeds and tool pin profiles (Elangovan et al., 2008). Pin profile plays a vital role in material mixing; and rotational speed significantly affects the heat input at welded zone. BM contains rolled structure of elongated grain that changes into equiaxed grains in welded region due to SPD in the processed zone

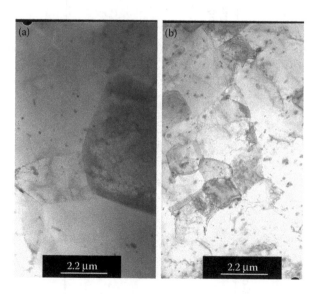

FIGURE 3.7   TEM micrographs of the SZ of joints welded by (a) conical tool and (b) scrolled tool. (Reprinted from *Materials and Design*, 30, N.D.M. Rodrigues et al. Influence of friction stir welding parameters on the microstructural and mechanical properties of AA 6016-T4 thin welds. 1913–1921, Copyright 2009, with permission from Elsevier.)

irrespective of rotational speed and type of pin profile. The main function of tool pin profile is to effectively move the plasticized material toward trailing side to complete the joint. Different pin profiles show different stirring action depending upon their profile features. The presence of flat faces in the triangular, square, and hexagonal shapes produces pulsating action during stirring of the softened material, whereas this phenomenon does not occur while using cylindrical, tapered, and threaded cylindrical pins. Pulsating action produced by pins having flat faces results in homogeneous redistribution of strengthening precipitates (Mg_2Si) and reduction in grain size (Elangovan et al., 2008). Rotational speed is one of the important FSW process parameters that significantly affect joint strength. Rotational speed directly controls the generation of heat at SZ. Higher rotational speed results in larger heat generation causing dissolution and coarsening of precipitates. Lower rotational speed, on the other hand, results in inadequate heat input leading to improper mixing of flowing material leading to formation of defects. The strength increases by increasing rotational speed up to a certain level beyond which it decreases. Adequate heat generation by optimum rotational speed and proper stirring of material by pulsating action of pin with flat faces results in superior mechanical properties with defect-free joints.

FSW of AA6082 was studied and the effect of welding speed (40–460 mm/min) on mechanical and microstructure of welded joints was investigated (Cavaliere et al., 2008). Variation in welding speed resulted in significant change in grain size. Fine and uniform distribution of grains in SZ was obtained on increasing the welding speed. Also, strength was found to increase by increasing welding speed up to 115 mm/min beyond which it decreased. Lower welding speed resulted in reduction in lowest hardness value, especially during welding of precipitate-hardened aluminum alloys (Lee et al., 2003a; Simar et al., 2008). Strength and hardness of FSWed Al–Mg–Si age-hardened alloys depend mainly on size, distribution, and volume fraction of precipitates in TMAZ and HAZ and it has been reported that FSW of age-hardened alloys resulted in the softening of weld region (Liu et al., 2003; Moreira et al., 2007). Also, FSW of artificially aged aluminum alloys resulted in dissolution of strengthening precipitates in SZ (Heinz and Skrotzki, 2002).

3.7 FSW OF 7XXX SERIES ALUMINUM ALLOYS

7xxx series are high-strength age-hardenable aluminum alloys that are widely used in aerospace structural components. FSW of some 7xxx series aluminum alloys is presented in Table 3.4. Precipitation sequence in Al–Zn–Mg–Cu alloys is given as

$$\alpha \,(\text{Supersaturated}) \rightarrow \text{G.P. zone} \rightarrow \eta' \rightarrow \eta \,(\text{Stable})$$

where η' is metastable phase and η is equilibrium phase. For copper-free alloy, η will be $MgZn_2$.

Grain refinement and tensile properties in FSW of AA7075-T6 was investigated and compared with BM (Goloborodko et al., 2004). BM contains elongated pancake-shaped grains after rolling and artificial aging with 100 nm fine precipitates in grain interiors and grain boundaries. Fine and equiaxed grains with average size of 3 μm are formed in SZ of welded joint. Grains are homogeneously distributed inside the SZ with around 90% of developed grain boundaries having high angle misorientations. Second-phase particles in grain interiors and boundaries with no dislocation substructure are formed in SZ. This shows that FSW can effectively produce fine-grained microstructure due to some mechanism similar to CDRX. CDRX through a dislocation-glide-assisted-subgrain rotation mechanism produced fine-grained structure during FSW as revealed by Jata and Semiatin (2000). Tensile testing of welded joint showed the possibility of improving toughness of 7xxx aluminum alloys by using FSW (Goloborodko et al., 2004).

FSW of AA7075-T6 was also investigated by researchers to understand the effect of FSW parameters on mechanical properties of welded joints (Rajakumar et al., 2011). Sound joints and also joints with defects are obtained at different combinations of process parameters and tool geometries. Defects are formed due to inadequate breaking of oxide layer at the interface caused by insufficient heat generation and improper material movement around the pin. Microhardness of SZ is found to be higher than that of BM, which may be due to refinement of grains based on Hall–Petch equation and the formation of IMCs. Higher values of microhardness in SZ compared to TMAZ and HAZ are attributed to finer grains in SZ, coarse-elongated grains in TMAZ, and coarse grains similar to BM in HAZ. Cai et al. (2005) also showed the refined grains in SZ, elongated and distorted grains in TMAZ, and grains similar to BM in HAZ. Similar variation in grains of welded region was observed by Kadlec et al. (2015) during FSW of AA7475. Figure 3.8 depicts hardness profile of welded joint as a function of grain size, which shows that microhardness closely depends on average grain size. Fine grains of 3–7 μm are observed in SZ

TABLE 3.4   FSW of 7xxx Series Aluminum Alloys

Material	Thickness (mm)	Rotation Speed (rpm)	Welding Speed (mm/min)	Pin Profile	References
7075-O	4.9	630	32	Threaded cylindrical	Dehghani et al. (2015)
7449-T7	3.2	1300	480	Threaded triflat	Wang et al. (2013)
7039-T6	5	410, 540, 635.	75, 120, 190	Threaded tapered	Sharma et al. (2012)
7020-T6	4	300, 600, 900, 1620	100, 500, 900	Cylindrical, tapered cylindrical with three flats	Lorrain et al. (2010)
7075-T6	3	1500	300	–	Goloborodko et al. (2004)
7475-T7351	6.35	400	400	–	Kadlec et al. (2015)
7020-T6	4	400, 600, 800, 1000	100	Cylindrical	Golezani et al. (2015)
7050-T7651	6.35	200, 600	109, 101.6	Cylindrical	Fuller et al. (2010)
7075-T6	3.175	640	160	–	Fuller et al. (2010)
7020-O	12	1120, 1400, 1800	20, 40, 80	Cylindrical	Gaafer et al. (2010)

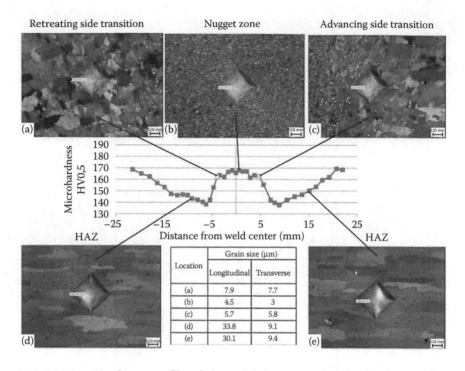

Retreating side transition   Nugget zone   Advancing side transition

Location	Grain size (μm)	
	Longitudinal	Transverse
(a)	7.9	7.7
(b)	4.5	3
(c)	5.7	5.8
(d)	33.8	9.1
(e)	30.1	9.4

FIGURE 3.8 Hardness profile of the welded region near the bottom surface with specific grain sizes. (Reprinted from *International Journal of Fatigue*, 74, M. Kadlec et al. Mechanical behaviour of AA 7475 friction stir welds with the kissing bond defect. 7–19, Copyright 2015, with permission from Elsevier.)

due to dynamic recrystallization (Figure 3.8b). Increase in grain size from SZ to HAZ results in reduction in microhardness accordingly (Figure 3.8). TMAZ contains elongated and partly recrystallized grains, which are larger in SZ and smaller in HAZ (Figure 3.8a and c). Zone experienced by heat alone and having grain similar to BM is characterized as HAZ (Figure 3.8d and e). Although grain size of HAZ is similar to that of BM, microhardness of BM is much higher than that of HAZ. Maximum hardness of BM may be attributed to its T-7351 condition with the presence of fine η' precipitates (Rajakumar et al., 2011). Similar observations are reported by Dehghani et al. (2015), that hardness decreases from SZ to HAZ and increases from HAZ to BM. Lowest hardness of HAZ is also reported by Leonard (2000) and Linton and Ripley (2008) during FSW of 7075-T6 due to coarsening of precipitates. Hardness of SZ is lower than that of BM due to partial dissolution of precipitate. Hardness profile of FSWed high-strength age-hardened aluminum alloys are strongly dependent on the distribution of precipitates (Genevois et al., 2005).

Sufficient amount of heat input with proper stirring is required to produce sound joints. Heat input and material flow during FSW are affected by rotational speed, welding speed, axial force, and tool geometry. Also, microstructure of FSWed region depends on the amount of heat input, degree of plastic deformation, and initial condition of BM. Lower heat input causes improper material mixing in SZ. Sharma et al. (2012) investigated FSW of AA7039 and revealed that onion ring and zig-zag line are observed in SZ at lower heat input, and higher heat input results in minimization/elimination of onion rings and zig-zag line. Heat input can be easily controlled by varying rotational speed. Increasing rotational speed increases heat input resulting in intense material mixing. Too high rotational speed, however, causes coarsening of grains at SZ and microvoids due to excessive release of stirred material leading to the reduction in microhardness and strength. Lower rotational speed generates inadequate heat resulting in improper material mixing. For a typical case, the joint strength reaches its maximum by increasing rotational speed from 900 to 1400 rpm and then starts to decrease with further increase in rotational speed (Rajakumar et al., 2011). Average grain size of SZ depends on the amount of heat input and strain rate. Higher rotational speed results in higher heat input and strain rate. Coarse grains are obtained at high heat input and higher deformation level results in finer grains. Golezani et al. (2015) investigated FSW of AA7020-T6 and found that FSW results in finer grains compared with BM and grain size increases with rotational speed as shown in Figure 3.9. They also revealed that heat input/peak temperature of the SZ (rather than deformation level) during FSW was the dominant factor in their study because grain size increased by increasing rotational speed (Figure 3.9), which in turn decreased microhardness at SZ according to Hall–Petch equation as shown in Figure 3.10. Too high heat input is responsible for larger grain size and accordingly lower microhardness at higher rotational speed.

Investigation on microstructure and second-phase particles of different zones was performed and compared with BM during FSW of AA7050-T651 (Su et al., 2003). TEM image of grain structures in BM and different weld zones are shown in Figure 3.11. BM contains pancake grains with subgrains of 1–5 μm in size (Figure 3.11a and b). Grains in HAZ are less affected by FSW and appear similar to that of BM (Figure 3.11c). Elongated grains with coarse subgrains and high dislocation density are found in TMAZ I (Figure 3.11d). Figure 3.11e shows TMAZ II region containing recovered fine subgrains with high density of subboundaries and

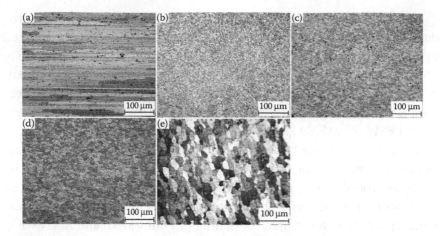

FIGURE 3.9   (a) Microstructure of BM, and microstructure of SZ of joints welded at different rotational speed of (b) 400, (c) 600, (d) 800, and (e) 1000 rpm. (With kind permission from Springer Science+Business Media: *International Journal of Advanced Manufacturing Technology*, Elucidating of tool rotational speed in friction stir welding of 7020-T6 aluminum alloy, 81, 2015, 1155–1164, A.S. Golezani et al.)

low dislocation density. SZ is characterized by fine equiaxed grains with size of 1–4 μm (Figure 3.11f) and varying dislocation density.

Fine distribution of granular precipitates and the presence of coarse precipitates in high- and low-angle boundaries are revealed by TEM

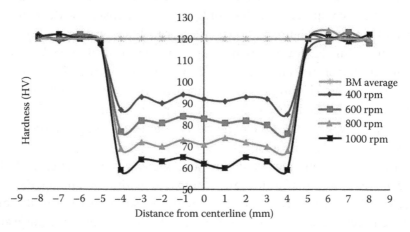

FIGURE 3.10   Hardness profiles of the joints welded at different rotational speeds. (With kind permission from Springer Science+Business Media: *International Journal of Advanced Manufacturing Technology*, Elucidating of tool rotational speed in friction stir welding of 7020-T6 aluminum alloy, 81, 2015, 1155–1164, A.S. Golezani et al.)

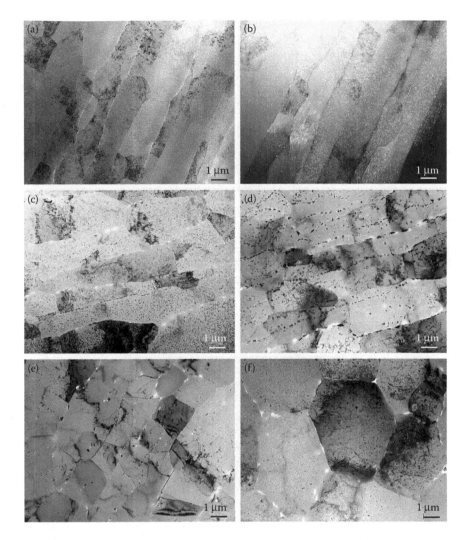

FIGURE 3.11 Grain structure of, (a) BM, (b) dark field image of region (a), (c) HAZ, (d) TMAZ I, (e) TMAZ II, and (f) SZ. (Reprinted from *Acta Materialia*, 51, J.Q. Su et al. Microstructural investigation of friction stir welded 7050-T651 aluminum, 713–729, Copyright 2003, with permission from Elsevier.)

micrographs of BM, as shown in Figure 3.12a. Precipitate-free zone (PFZ) having 25 nm width is found along the grain boundary of BM (Figure 3.12a). Figure 3.12b shows homogeneously distributed coarse precipitates in HAZ with significantly increased PFZ. Intragranular precipitates (η' and η) are present in HAZ, and even the peak temperature in HAZ (250°C) is higher than precipitate (η) dissolution temperature (180–260°C). This may be due

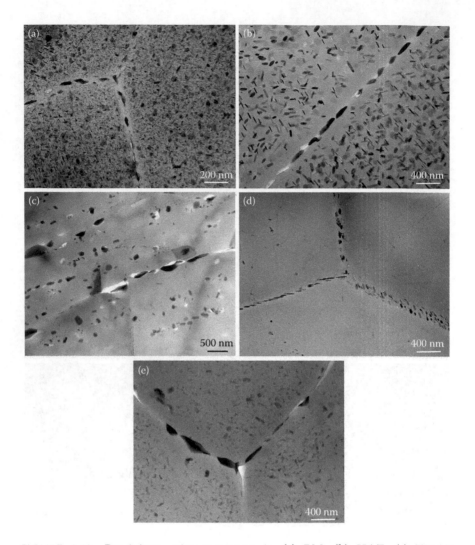

FIGURE 3.12 Precipitate microstructures in (a) BM, (b) HAZ, (c) TMAZ
I, (d) TMAZ II, and (e) SZ. (Reprinted from *Acta Materialia*, 51, J.Q. Su et al.
Microstructural investigation of friction stir welded 7050-T651 aluminum,
713–729, Copyright 2003, with permission from Elsevier.)

to inadequate holding time at dissolution temperature of precipitates (Su
et al., 2003).

TMAZ I region shows duplex population of precipitates with coarse
one due to higher peak temperature (>350°C) and fine reprecipitated
particles (Figure 3.12c). In TMAZ II region, the precipitates are fully
dissolved due to attainment of the solution heat-treatment temperature
and reprecipitation took place along the grain boundaries, subgrain

boundaries, and on dislocations during cooling (Figure 3.12d). In SZ, precipitates dissolve and reprecipitation takes place on dislocation (Figure 3.12e). Uniform dispersion of precipitates of 60–100 nm in size is observed throughout the matrix (Su et al., 2003).

TEM image of BM (AA7075-T651) shows duplex populations of precipitates with one group of 50–75 nm and other group of 10–20 nm (Figure 3.13a). Figure 3.13b shows disk or plate-type randomly oriented intragranular precipitates in SZ (Rao et al., 2015). It is revealed from Figure 3.13b that a portion of precipitates are dissolved and reprecipitated; whereas, some precipitates get coarsened indicating attainment of temperature higher than the solution temperature of precipitates in the SZ with insufficient time for dissolving all precipitates. TEM image of TMAZ (Figure 3.13c)

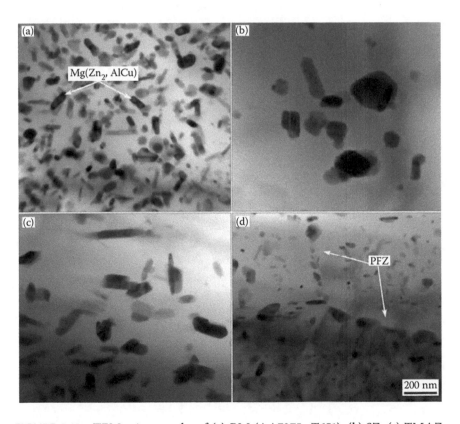

FIGURE 3.13   TEM micrographs of (a) BM (AA7075–T651), (b) SZ, (c) TMAZ, and (d) HAZ. (Reprinted from *Transactions of Nonferrous Metals Society of China*, 25, T.S. Rao et al., Microstructure and mechanical properties of friction stir welded AA7075–T651 aluminum alloy thick plates, 1770–1778, Copyright 2015, with permission from Elsevier.)

shows that larger hardening precipitates are not significantly affected during welding whereas, smaller precipitates become coarser. Figure 3.13d shows TEM micrograph of HAZ containing homogeneously distributed intragranular precipitates and grain boundary precipitates with PFZ surrounding grain boundaries.

CHAPTER 4

Friction Stir Welding of Dissimilar Aluminum Alloys

4.1 INTRODUCTION

Welding of dissimilar metal has always been a challenging task during fusion welding. The problem is mainly caused by lack of chemical, metallurgical, and all other compatibilities. During dissimilar fusion welding, the progress of solidification in weld metal witnesses a constant segregation along temperature gradient because of rejection of atoms due to lack of solubility. Moreover, in several dissimilar systems, the subsequent cooling results in segregation in liquid phase itself. Joint consolidation in fusion welding is greatly influenced by system equilibrium and kinetics and thus in most systems' fusion of dissimilar welding is unsuccessful. In systems where dissimilar welding is possible, the region near fusion boundary is significantly different from the bulk material. In fact, it typically comprises five characteristically different zones: (i) unmixed zone, (ii) filler-metal-depleted zone, (iii) partially mixed zone, (iv) intermediate mixed zone, and (v) hard zone (Savage et al., 1976). In any case, in very limited system where dissimilar welding is possible, elemental integrity and homogeneity across the joint does not exist.

Since solid state welding process joint consolidation is not derived from metallurgical equilibrium and system kinetics, there exists a great potential for dissimilar welding. In solid state joining processes, oxide films and surface contaminants at the interface are expelled from the interface by plastic deformation. The interface is heated locally by interfacial slipping due to plastic work, which causes temperature rise and softening of the materials. This also brings the clean and activated metallic interface in close contact, forming bond and creating the joint. Thus, theoretically, this process has enormous possibilities in joining metals combinations irrespective of their metallurgical compatibility. FSW, being a solid state welding process, can effectively join dissimilar aluminum alloys by SPD with low heat input compared to fusion welding resulting in less loss of prewelding processing advantages and improved joint strength.

4.2 ISSUES WITH DISSIMILAR MATERIALS WELDING

The issue of welding of dissimilar aluminum alloys is very complex as it depends on the type of alloys being welded. Different aluminum alloys contain specific alloying elements wherein the concentration of these elements varies widely leading to large variation in properties specially the melting point and thermal conductivity. Relatively lesser concentration of alloying elements results in high melting point and thermal conductivity. Welding of alloys with higher thermal conductivity coupled with higher melting point requires high heat input at the weld and vice versa. Problems in welding of dissimilar aluminum alloys arise primarily due to large difference in thermal properties of the alloys being welded. During usual fusion welding of alloys with distant thermal properties, there exists vast difference in fluidity, superheat, and solidification rate on each side of alloy. Generally, most common problems related to dissimilar aluminum alloy welding can be grouped as (i) hydrogen embrittlement, (ii) hot cracking, and (iii) formation of IMCs. The most critical problems arising from this is variation in solubility of hydrogen with temperature, which is mainly responsible for hydrogen embrittlement. Further, because of variation in thermal conduction coefficient and melting point, weld of these alloy combinations are also prone to hot cracking; mainly liquidation cracking in HAZ and solidification cracking in the weldment. The various metallic atoms present as alloying elements result in the formation of hard and brittle IMCs. During welding of dissimilar alloys of same or different series, formation of IMCs makes the joint weak and brittle. All these problems cause difficulty in the welding of dissimilar aluminum alloys, especially using fusion welding techniques. Most of these issues are caused by the change of state during fusion welding and therefore, solid state welding processes have gained considerable success as it does not involve melting and solidification.

4.3 MAJOR CHALLENGES IN THE FSW OF DISSIMILAR MATERIALS

FSW has evolved as an ideal maiden process, which has shown remarkable success in joining dissimilar alloys in general and dissimilar aluminum alloy combinations in particular. It is because of this reason that during FSW joint consolidation is obtained by mechanical mixing of softened material and it does not require liquid state dissolution. The merit of this process is that suitable choice of some of its parameters such as tool offset, welding position, etc. offers more flexibility to take care of variation in material properties. During welding, the stirring action of the tool moves softened material from AS and mixes the materials from both plates before it is finally deposited through the RS to the back of the tool. The flow stress involved in the movement of the material mainly depends on the strength of alloy, its melting point, softening temperature, and the temperature prevailing during welding.

Although FSW is regarded as a favorite method for dissimilar joining, it is imperative to mention that it cannot be readily used in its simple form to weld dissimilar welding, rather it needs suitable adaptations to take care of variations in properties of softened material on both sides. As the FSW tool produces specified conditions of heat and temperature, depending on the FSW process parameter, this condition will equally prevail on both side. At same conditions of temperature, two materials of different mechanical properties witness different softening and consequently flow stress in both the materials is different. In this situation, material with lower melting point will be softer as compared to the material with higher melting point. This unequal softening at the same temperature will cause different flow stress in both materials and there will be an imbalance in material movement and joint consolidation will not occur. Thus, to address issues related to welding of dissimilar material, two important process parameters, that is, tool offset and side of material should be tactfully used so that flow stress balance is struck and joint consolidation is obtained.

4.4 JOINING OF DISSIMILAR ALUMINUM ALLOYS

The common techniques used for joining dissimilar aluminum alloys are mechanical fastening, welding, soldering, brazing, adhesive bonding, etc. Among all the techniques for joining dissimilar aluminum alloys, welding is an efficient joining process with various advantages in terms of flexibility, cost, productivity, joint efficiency, stiffness, etc. Welding processes including both fusion and solid state are used for joining dissimilar aluminum alloys. However, fusion welding processes pose various problems associated with melting and solidification of material being welded and selection of filler wire and other consumables compatible for both the materials.

Further, joint strength of fusion welds is usually lower than those in solid state processes due to (i) higher temperature generated in fusion welding than that of solid state welding and an associated microstructure of solidified and heat-affected regions and (ii) difference in properties of filler material with that of both parent materials. Solid state welding is free from problems associated with fusion welding and has great opportunities in joining dissimilar aluminum alloys though they have their own problems. Solid state welding processes include ultrasonic welding, diffusion welding, explosion welding, FW, and FSW. FSW with optimum parameter combination is found to be an efficient joining process for joining any combination of dissimilar aluminum alloys. Earlier, riveting was used for joining high-strength aerospace aluminum alloys (2xxx and 7xxx series aluminum alloys) since these were not suitably welded by conventional fusion welding processes. Now FSW, being a solid state welding process, has replaced riveting and has proved to be an effective method for joining high-strength aluminum alloys and has also resulted in huge cost and weight saving.

High-strength aluminum alloys are widely used in shipbuilding, automotive, aerospace, and railway industries. For instance, 5xxx series aluminum alloys are used in automobiles in inner part of door panels and 6xxx series aluminum alloys are used in outer part of door panels. This requires the dissimilar joining of both series of aluminum alloy in door panels. Similarly, outer hulls of ships are made by 5xxx series aluminum alloys possessing good corrosion resistance for seawater while inner structural parts of shipbuilding are made of 6xxx series aluminum alloys possessing good appearance and strength. This gives rise for dissimilar joining of both series of aluminum alloy. In aircraft structures, lower portion and upper portion of wings are made of 2xxx series and 7xxx series aluminum alloys, respectively, which require dissimilar joining of these aluminum alloys.

4.5 FSW OF DIFFERENT ALLOYS

Manufacturing and transportation industries have shown interest in joining of dissimilar aluminum alloys, but problems associated with difference in properties of both the materials make this difficult. Efficient welding of dissimilar material is a challenging task for engineers and scientists due to difference in their mechanical, physical, thermal, and metallurgical properties. FSW is found to be suitable for joining dissimilar aluminum alloys as it is performed by frictional heating and SPD of materials being joined without melting the parent materials. SPD causes dynamic recrystallization in SZ resulting in fine equiaxed grains with good mechanical properties.

High-strength aluminum alloys possess desirable properties that are required in engineering structures. Its high-strength-to-weight ratio is highly desirable in transportation industries. Therefore, they are widely used in fabrication of automotive vehicle structures, shipbuilding, aircraft structures, and other structural applications. Dissimilar welding of aluminum alloys is employed in various industries where components and parts are required to sustain complex loading and different environmental conditions when put to use. Lower heat input in FSW makes it suitable for joining dissimilar aluminum alloys, which conserves the prewelding processing of alloys (heat treatment, strain hardening, etc.). Also, problem associated with solidification cracking that significantly degrades the mechanical properties during fusion welding is overcome by using FSW.

Although FSW can efficiently join dissimilar aluminum alloys, improper selection of process parameters and welding conditions could degrade the joint quality. Along with other FSW process parameters, tool pin offset (Cole et al., 2014; Khan et al., 2015b) and material position (Guo et al., 2014) significantly affects the mechanical properties and microstructure of the welded joint. FSW of dissimilar aluminum alloys has attracted extensive research interest due to unavailability of suitable conventional fusion welding processes, which has led to the increased importance of this process in engineering applications. Dissimilar aluminum alloys have been welded by using FSW with excellent joint efficiencies (Luijendijk, 2000; Khodir and Shibayanagi, 2008; Cavaliere et al., 2009).

4.6 FSW OF 5XXX–6XXX SERIES ALUMINUM ALLOYS

Due to large application of 5xxx and 6xxx combination, there has been a growing interest in the welding of these alloys. FSW of dissimilar combination of 5xxx–6xxx series aluminum alloys is presented in Table 4.1. Dissimilar FSW of AA5086 and AA6061 was studied at different rotational speed (600–1200 rpm) and welding speed (100–150 mm/min) (Aval et al., 2011). Tunnel-type defects were observed in the joints fabricated with low rotational speed due to insufficient material transport and low

TABLE 4.1 FSW of Dissimilar Combination of 5xxx–6xxx Series Aluminum Alloys

Material	Thickness (mm)	Rotation Speed (rpm)	Welding Speed (mm/min)	Pin Profile	References
6061-T6 and 5086-O (variable)	5	600, 840, 900, 1000, 1200	100, 150	TC	Aval et al. (2011)
5182-H111 (AS) and 6016-T4	1	1120	320	ThC	Leitao et al. (2009)
6351-T6 (AS) and 5083-H111	6	600, 950, 1300	60	SS, SH, SO, TS, TO	Palanivel et al. (2012)
AA6351-T6 (AS) and 5083 H111	6	600, 775, 950, 1125, 1300	36, 49.5, 63, 76.5, 90	SS, TS, SH, SO, TO	Palanivel et al. (2013)
6082-T6 (AS) and 5754-H111	5	800, 1000, 1250	63, 80, 100	ThC	Kasman (2013)
5083-H111 and 6082-T651	6	1250	64	ThTC	Gungor et al. (2014)
6351-T6 (AS) and 5083-H111	6	950	36, 63, 90	SS	Palanivel et al. (2014)
5052-H32 and 6061-T6	2	2000	100	SC	Park et al. (2010)
6061 (AS) and 5083	3	890, 1540	118, 155	SC	Shigematsu et al. (2003)
5083-H116 (AS) and 6063-T6	4.75	450	100	SC	Khan et al. (2015b)
6061(AS) and 5052	1 and 1.5	1500	63, 98	SC	Doley and Kore (2016)
5083 and 6082 (variable)	3	280, 560, 840	100, 200, 300	ThP	Steuwer et al. (2006)
7475-T761 and 2219-O	2.5	900	100	ThC	Khan et al. (2017)

SC—straight cylinder; ThC—threaded cylinder; ThTC—threaded tapered cylinder; SS—straight square; SH—straight hexagon; SO—straight octagon; TS—tapered square; TO—tapered octagon; TC—tapered cylinder; ThP—threaded pentagon.

heat generation. High rotational speed also resulted in defects due to turbulent flow of material in SZ. Therefore, in order to produce sound joints a careful selection of optimum rotational speed is needed to facilitate generation of adequate amount of heat input and appropriate flow of material. Mixing of dissimilar aluminum alloys in SZ at different material position and rotational and welding speed is shown in Figure 4.1. Mixing of material improves by increasing rotational speed and decreasing welding

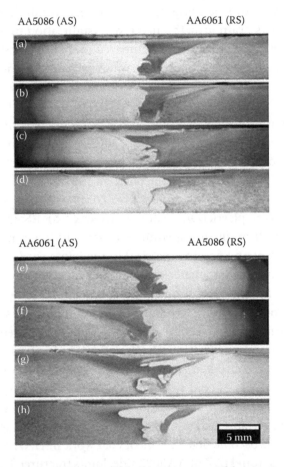

FIGURE 4.1 Macrostructure of welded joints at variable rotational and welding speed of (a) 840 rpm, 100 mm/min, (b) 840 rpm, 150 mm/min, (c) 900 rpm, 100 mm/min, (d) 900 rpm, 150 mm/min, (e) 1000 rpm, 100 mm/min, (f) 1000 rpm, 150 mm/min, (g) 1200 rpm, 100 mm/min, and (h) 1200 rpm, 150 mm/min. (With kind permission from Springer Science+Business Media: Thermo-mechanical and microstructural issues in dissimilar friction stir welding of AA5086–AA6061, *Journal of Materials Science*, 46, 2011, 3258–3268, H.J. Aval et al.)

speed in a particular range. Effective mixing of material was obtained in SZ when AA5086 was located at AS. Adequate flow of plasticized material was disturbed on placing AA5086 at RS due to lower temperature at RS, and material flow readily occurred when AA5086 was located in AS due to higher temperature in this region (Aval et al., 2011). Higher flow stress of AA5086 requires higher temperature for proper mixing, which favors the placement of AA5086 at AS. Temperature at AS was higher due to high strain rate during stirring. In dissimilar FS weld, finer grains in SZ were produced at AA6061 side compared to AA5086. Second-phase particles in AA6061 restricts grain growth during recrystallization in SZ at AA6061 side and results in smaller grains compared to that in AA5086 side. This also results in higher microhardness in SZ at AA6061 side than that at AA5086 side.

In FSW of AA5083-H111 and AA6351-T6, different tool rotational speeds and pin profiles were used to investigate the flow of material and mechanical properties (Palanivel et al., 2012). Variation in material flow behavior was observed at different tool pin profiles. Contact area between plasticized material and tool pin depend on the pin profile. Straight tool pin has larger and uniform contact area compared to tapered pin, and transportation of plasticized material from AS to RS throughout the thickness is uniform during welding with straight tool pin. Less interaction between tool and flowing material in the joints produced by tapered pin resulted in irregular material flow and reduction in frictional heat leading to the formation of tunneling defect behind the tool pin. Straight pin profile produced defect-free joints with higher tensile strength. Joint fabricated using straight square pin with rotational speed of 950 rpm yielded maximum tensile strength. Variation in tensile strength in dissimilar joints was observed, which may be attributed to degree of plastic deformation, loss of cold work in strain-hardened alloy, the presence of macroscopic defects, and dissolution and coarsening of strengthening precipitate in age-hardened alloy. Failure of joints from SZ was due to the presence of macroscopic defect. Loss of cold work in HAZ resulted in the failure of joints from HAZ of AA5083 side. Joints fractured from TMAZ of AA6351 side was attributed to the dissolution and coarsening of strengthening precipitates. Different flow stress possessed by both the aluminum alloys resulted in vortex-like flow pattern during stirring. Lamellae-like structure was formed in the mixed flow region of welded joints, which was attributed to intense plastic deformation and layer-by-layer movement of plasticized material (Palanivel et al., 2012). Leal et al. (2008) investigated

FSW of dissimilar aluminum alloys (AA5182 and AA6061) and reported that onion rings with intercalated bright (AA6061) and dark (AA5182) layers were formed at rotation speed of 1800 rpm, welding speed of 160 mm/min, and tool tilt angle of 2.5° using threaded pin profiled tool with concave shoulder.

Material mixing during FSW of dissimilar aluminum alloys (AA5052-H32 and AA6061-T6) was investigated by Park et al. (2010). They reported that the fixed position of alloys played a significant role in material mixing of dissimilar alloy at SZ. Onion rings with dark (AA5052-H32) and bright (AA6061-T6) regions were observed at rotational speed of 2000 rpm, 100 mm/min, and tool tilt angle of 3°. Significant difference in material mixing occurred by changing the position of materials from AS to RS, which resulted in different mechanical properties and microhardness distribution (Park et al., 2010). Macrostructure of dissimilar FS-welded AA5083 and AA6061 alloy is shown in Figure 4.2 (Shigematsu et al., 2003). Figure 4.2a and b indicates the cross section perpendicular and parallel to the welding direction, respectively. Material flow of both alloys was clearly visible in SZ due to different etching response. Vortex-type layered structure was observed at lower portion of SZ due to intense plastic deformation of both alloys by threaded pin.

FS-weld fabricated at lower rotational speed (600 rpm) resulted in lower joint strength. Tensile strength increased by increasing rotational speed to 950 rpm and then decreased subsequently (Palanivel et al., 2013). Lower strength obtained in the joints produced at low heat input condition (low rotational speed) was due to lack of material mixing in SZ. In case of higher rotational speed, large heat was generated, which resulted in turbulent flow of material around the tool pin. Further, high rotational speed generated more amount of flash coming out of the tool shoulder during FSW, which resulted in the formation of voids (Elangovan and Balasubramanian, 2007).

Welding speed affects the heat input, which in turn affects the tensile strength during FSW of dissimilar aluminum alloys. Lower strength was obtained at lower welding speed due to longer exposure time with high heat input leading to turbulent flow of material (Palanivel et al., 2013). Tensile strength increased by increasing welding speed and then decreased after reaching its maximum value (Colligan et al., 2003). Lower tensile strength at higher welding speed is attributed to short exposure time with inadequate heat input and inappropriate material movement leading to the formation of void and defects like KB.

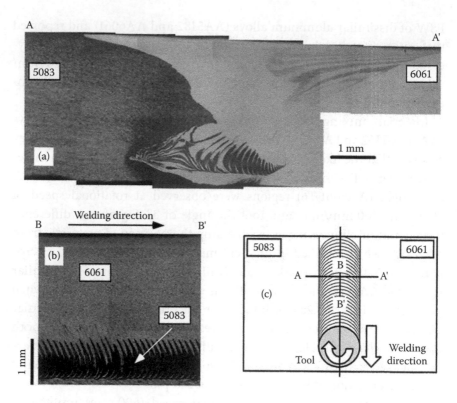

FIGURE 4.2 Cross sections of dissimilar FS weld of AA5083 and AA6061: (a) perpendicular to the welding direction (line indicated by A–A in (c)) and (b) parallel to the welding direction (line indicated by B–B in (c)). (With kind permission from Springer Science+Business Media: Joining of 5083 and 6061 aluminum alloys by friction stir welding, *Journal of Materials Science Letters*, 22(5), 2003, 353–356, I. Shigematsu et al.)

4.7 FSW OF 2XXX–7XXX SERIES ALUMINUM

Aerospace grade aluminum alloys (AA2xxx and AA7xxx) are high-strength heat-treatable alloys. Generally, 7xxx series aluminum alloys are used in aircraft fittings, missile parts, worm gears, structural components, other aerospace, as well as defense applications. Furthermore, 2xxx series aluminum alloys are mostly used in aircraft fittings, gears and shafts, bolts, couplings, missile parts, fuel tanks, pistons, rectifier parts, worm gears, and structures. It is clear that there are several common applications of these two series of aluminum alloys. Consequently, joining of such alloys is required in many applications. Therefore, an efficient joining process is required to join these high-strength aluminum alloys. Unfortunately, these alloys show poor weldability with conventional welding processes due to various problems associated with their solidification microstructure. Dissimilar welding of these alloys is even more difficult due to large difference in various properties. Solid state welding techniques can overcome problems associated with fusion welding. The demand for efficient joining process of dissimilar aluminum alloys by aerospace industries is successfully met by FSW process and that is why it is widely used by aircraft manufacturers.

Understanding of joining high-strength aluminum alloys by FSW is important for enhancing their use in various applications. Joint quality of FSW welds considerably depends on the flow of material and mixing. Material flow in FSW of dissimilar aluminum alloy is much more complex as compared to similar metal welding and, therefore, it needs to be clearly addressed and understood. Many researchers considered various factors and weld conditions and also investigated the effect of different process parameters on joint quality and material flow of dissimilar aluminum alloys for better understanding of the involved phenomena. Dissimilar FSW of 2xxx–7xxx series aluminum alloys is presented in Table 4.2. Amancio-Filho et al. (2008) found that excessive thermal softening during FSW of aluminum alloys is responsible for poor strength of the welded joints. Sato et al. (2006) studied microstructural behavior of dissimilar FS-welded high-strength aluminum alloys and reported that material flow around the tool pin was significantly affected by the amount of heat input during welding and different microstructural features were observed at varying heat input. Khan et al. (2017) investigated the mechanical and microstructural features of similar and dissimilar FSW of AA7475 and AA2219 alloys and reported that border line was visible at the interface of both the alloys due to improper material mixing but joint strength was not affected by the mere presence of boundary line at the interface.

In FSW of 2024–7075, fine equiaxed grains were obtained due to SPD at SZ (Silva et al., 2011). Sharp transition of TMAZ/HAZ was observed at AS, whereas this transition was not as much clear at RS due to comparatively higher strain rate and severe deformation taking place at AS. Microstructural evolution of dissimilar FSW joints from no-mix condition at lower rotational speed to intimate-mixing condition obtained at higher rotational speed is shown in Figure 4.3a and b. Clear border between AA2024 and AA7075 is visible in SZ at lower rotational speed due to inadequate mixing of dissimilar materials. This may be due to excessive movement of bulk material from AS to RS under the shoulder without adequate softening. Border between the two alloys is easily observed, which

TABLE 4.2 Dissimilar FSW of 2xxx–7xxx Series Aluminum Alloys

Material	Thickness (mm)	Rotation Speed (rpm)	Welding Speed (mm/min)	Pin Profile	References
2024-T3 and 7075-T6	2.5	–	160	Cylindrical	Cavaliere et al. (2006)
2024-T3 and 7075-T6 interchange	3	400, 800, 1200, 1600, 2000	100	Threaded cylindrical	Khodir and Shibayanagi (2007)
2024-T3 (AS) and 7075-T6	4	1600	120	Threaded cylindrical	Cavaliere and Panella (2008)
7075-T6 (AS) and 2024-T3	3	400, 1000, 2000	254	Threaded cylindrical	Silva et al. (2011)
7075-O (AS) and 2024-T4	5	500, 800, 1,100	50 and 80	Threaded 4-flute and threaded taper	Bahemmat et al. (2012)
2219 (AS) and 7039	6	500, 710, 1000	20, 40	Threaded cylindrical	Venkateswarlu et al. (2015)
7475-T761 (AS) and 2219-O	2.5	900	100	Threaded cylindrical	Khan et al. (2017)
2024-T3 and 7075-T6	1.2, 2, 2.5	400,1200, 1500	100, 150,400	–	Zadpoor et al. (2010)
2024-T3 and 7075-T6	3	1200	42, 72, 198	Threaded cylindrical	Khodir and Shibayanagi (2008)

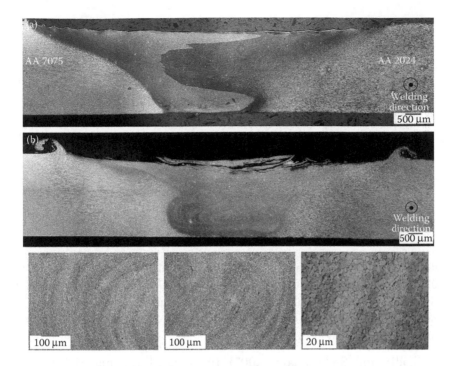

FIGURE 4.3 Macrostructure of FS-welded AA2024–AA7075 at rotational speed of (a) 400 rpm and (b) 2000 rpm. Details in (b) shows the onion ring in the middle of SZ and in SZ/TMAZ interface at AS and RS. (Reprinted from *Materials and Design*, 32, A.A.M.d. Silva et al., Material flow and mechanical behaviour of dissimilar AA2024-T3 and AA7075-T6 aluminum alloys friction stir welds, 2021–2027, Copyright 2011, with permission from Elsevier.)

is bent toward either side of the original faying line due to stirring action of the pin acquiring a continuous S shape (Figure 4.3a). Onion rings are observed in SZ of joint fabricated at higher rotational speed of 2000 rpm (Figure 4.3b). The same behavior of dependence of material mixing on rotational speed has also been reported by Sato et al. (2006) and Khodir and Shibayanagi (2007).

Microhardness profile of dissimilar joint between AA2024 and AA7075 obtained at rotational speed of 1000 rpm is depicted in Figure 4.4, which shows that hardness is minimum at HAZ of softer material (RS) side due to coarsening of precipitates of naturally aged aluminum alloy (2024-T3) (Silva et al., 2011). Heterogeneous distribution of microhardness is observed in SZ at different thickness. Abrupt change in microhardness at SZ clearly indicated the absence of intimate mixing between the alloys and the presence of boundary between dissimilar materials in SZ.

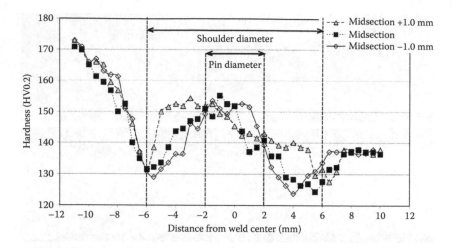

FIGURE 4.4 Microhardness distributions of dissimilar FS-welded joint (2024–7075) at various positions. (Reprinted from *Materials and Design*, 32, A.A.M.d. Silva et al., Material flow and mechanical behaviour of dissimilar AA2024-T3 and AA7075-T6 aluminum alloys friction stir welds, 2021–2027, Copyright 2011, with permission from Elsevier.)

Khodir and Shibayanagi (2007) investigated the material mixing behavior in SZ of dissimilar FS-welded heat-treatable aluminum alloys (AA2024–AA7075). They observed that mixing did not occur at lower rotational speed (400 rpm) leading to appearance of border between both the alloys in SZ. Onion rings were observed in SZ at higher rotational speed (800, 1200, and 1600 rpm) manifesting effective mixing of the materials. Clear bands in SZ were visible at AS due to higher strain rate and SPD compared to RS (Khodir and Shibayanagi, 2007). In dissimilar FSW of age-hardenable aluminum alloys, higher rotational speed resulted in unclear onion ring with irregular SZ shape. Size of SZ increased with the increase in rotational speed till 1200 rpm due to increasing heat input. Further, increase in rotational speed decreased SZ area, which may be attributed to excessive flash removal from SZ at higher rotational speed. Increasing rotational speed resulted in higher grain size in SZ and more homogenous distribution of fine precipitates. Grain size increased from 3.1 to 5.6 μm on increasing rotational speed from 400 to 2000 mm/min (Khodir and Shibayanagi, 2007). Similar results of grain coarsening with increase in rotational speed during FSW of AA2024 and AA6061 were also reported by Li et al. (1999b).

Dissimilar FS-welded AA2024 and AA7075 joint fabricated with rotational speed of 400 rpm showed clear borderline in SZ differentiating

AA2024 and AA7075 sides. Energy dispersive spectroscopy (EDS) analysis of SZ showed Cu-enriched region at AA2024 side and Zn-enriched region at AA7075 side as Cu and Zn are major alloying elements in AA2024 and AA7075, respectively (Khodir and Shibayanagi, 2007). From the border present in SZ toward AA2024 side, Cu content fluctuated from 5.1 to 6.3 mass % and decreased drastically just after crossing the border from 1.8% to 2.9%. Zn content increased sharply after crossing the border from 0.2% to 5.7%–8.5% toward AA7075 side. This showed the absence of proper mixing of both the alloys in SZ at low heat input condition (i.e., lowest rotational speed) leading to abrupt change of microhardness in SZ, while moving across the border. Microhardness of the joints fabricated with 1200 rpm increased due to proper mixing of materials and fluctuated due to the presence of alternate bands of onion ring having composition of both the alloys.

Tensile strength increased with increase in rotational speed up to a particular value beyond which it decreased. Defect-free joints fractured from HAZ of weaker material and in joint that failed in SZ showed the presence of KB or crack-type defects. Higher tensile strength in the longitudinal side was obtained for joints in which stronger material was fixed at AS. SZ was mainly occupied by the material located at AS. Therefore, tensile strength of SZ depends on fixed location of material. Transverse tensile strength showed higher values when softer material was located toward AS, whereas longitudinal tensile strength showed higher values when stronger material was located toward AS (Khodir and Shibayanagi, 2007).

Mechanical and microstructural behavior of dissimilar FS-welded AA7075-O and AA2024-T4 alloy was investigated (Bahemmat et al., 2012). Results revealed that the quality of welds of these dissimilar aluminum alloys depend on the amount of heat input, type of heat distribution, and degree of plastic deformation, which may result from combination of rotational speed, welding speed, and tool pin profile. In other words, formation of weld in FSW is governed by temperature in SZ, stirring action of the tool pin and forging action of the tool shoulder (Bahemmat et al., 2012). Joint will be defective if any (or some/all) of these factor(s) is/are inappropriately chosen. Different joints were produced at varying rotational and welding speed with threaded taper and 4-flute-threaded tools. Overall hardness of SZ was reduced with increase in rotational speed and decreasing welding speed. Four-flute tool resulted in lower hardness values compared to tapered threaded tool. However, grain size of the joints produced by 4-flute tool was smaller than that produced by tapered threaded tool. Therefore, in dissimilar joining of age-hardenable aluminum alloys, grain size does not significantly affect the microhardness distribution.

During tensile test, joints have defects fractured from SZ while defect-free joints showed failure from AA7075-O side, which is the weaker alloy. Mechanical properties of dissimilar welded defect-free joints are determined by the properties of weaker material. The macrostructures of dissimilar welds fabricated by different combination of rotational speed, welding speed, and tool pin profile is depicted in Figure 4.5.

It is evident from Figure 4.5 that joints fabricated with tapered threaded tool at rotational speed of 500 rpm for both welding speeds (50 and 80 mm/min) resulted in inadequate plasticization of material. Some defects remained at interface of the two alloys around the root of weld due to inadequate softening of material caused by low rotational speed, which led to improper stirring of material at the bottom. This defect is termed as *KB* or *JLR* and is caused by partial or no breaking of oxide film at the interface (Khan et al. 2015b). Similar defect was also observed by Oosterkamp et al. (2004) and they suggested that insufficient breakup of oxide layers at the faying surface of BMs is responsible for KB formation. It can be minimized by increasing rotational speed and plunge depth. Detailed discussion on KB defect can be found in the work of Khan et al. (2015b).

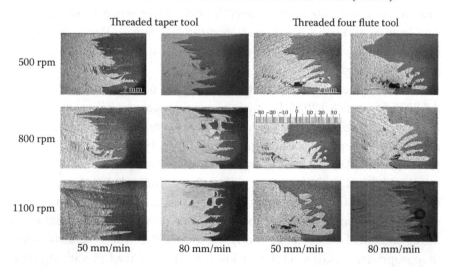

FIGURE 4.5 Macrostructure of weld cross section for different rotational and welding speeds and tool pin profiles. (With kind permission from Springer Science+Business Media: Study on dissimilar friction stir butt welding of AA7075-O and AA2024-T4 considering the manufacturing limitation, *International Journal of Advanced Manufacturing Technology*, 59, 2012, 939–953, P. Bahemmat et al.)

The majority of joints produced with threaded 4-flute tool showed inappropriate stirring of material (Figure 4.5). This may be attributed to the absence of vertical flow of material by using threaded 4-flute tool. Voids were observed in high rotational speed condition due to turbulent flow in SZ and can be prevented by reducing rotational speed. Both high and low rotational speed resulted in defect formation as less plasticized material, which is strong, resisted stirring and forging while overheated material, which is more soft, generated turbulent flow in SZ. According to Figure 4.5, all defects are located at AS (AA7075-O side) in welds fabricated by threaded 4-flute while defects are observed at interface in threaded tapered tool case.

During FSW of dissimilar aluminum alloys, higher peak temperature was obtained at 2024-T4 side compared to 7075-O side at a distance 10 mm from weld center (Bahemmat et al., 2012). The amount of heat required to raise a given degree of temperature is higher in 7075-O alloy compared to that in 2024-T4 due to higher heat capacity of AA7075-O ($cp = 0.96$ J/gC) compared to AA2024-T4 ($cp = 0.875$ J/gC) resulting in lower temperature generated at AA7075-O. Also, temperature generated at SZ depends on the plastic work done by the tool on material. Less plastic work was done by the tool on AA7075-O as it is softer than 2024-T4, which resulted in less temperature generated at AA7075 side (Bahemmat et al., 2012). Variation in temperature at different zones in welded joint significantly affects the microhardness profile as both AA2024 and AA7075 are heat-treatable alloys. Higher temperature generated in SZ due to higher strain compared to other zones leads to dynamic recrystallization and simultaneously dissolution and coarsening of strengthening precipitates. Precipitate dissolution and coarsening occurred in the SZ, which strongly affected the microhardness distribution and resulted in lower hardness than in BM. Hardness reduction in HAZ is attributed to over-aging and it is significantly affected by the thermal cycle during FSW of age-hardenable aluminum alloy. Thermal cycle experienced by HAZ can be controlled by adjusting rotational and welding speed. Dissimilar FS-welded defect-free joints of age-hardneable aluminum alloys fail from the region of minimum hardness while majority of defected joints fail from SZ due to stress concentration.

Dissimilar FSW of aerospace grade aluminum alloy (AA2219-O and AA7475-T761) was performed and defect-free joints were obtained at 900 rpm and 100 mm/min with threaded cylindrical pin profile (Khan et al., 2017). Figure 4.6 shows the microstructures of different zones of

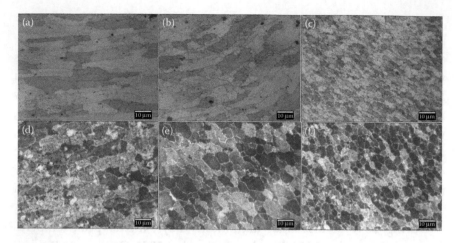

FIGURE 4.6 Microstructures of different zones of 2219–7475 welded joint: (a) HAZ of 7475 (AS), (b) TMAZ of 7475, (c) SZ of 7475, (d) HAZ of 2219 (RS), (e) TMAZ of 2219, and (f) SZ of 2219.

dissimilar joint at AS and RS. It is evident from Figure 4.6 that different microstructural zones such as SZ, TMAZ, and HAZ at AS and RS can be clearly identified due to difference in grain size and their orientation.

SZ experienced the highest strain among all weld zones and had undergone dynamic recrystallization, which resulted in grain refinement. Grain size of stirred AA7475 side was smaller than that of AA2219 side (Figure 4.6c and f) due to higher strain rate experienced by the material at AS compared with RS, which resulted in comparable higher grain refinement. Grains at AS faced higher stress due to sticking of material with the tool pin. Also, the presence of second-phase particle in AA7475-T761 restricted the movement of grain boundaries that resulted in lower grain size during recrystallization and such mechanism of grain growth restriction was absent in AA2219-O due to its annealed condition. Consequently, more refined grains were observed in AA7475-T761 side compared to that in AA2219-O side. Downward flow of material was observed in SZ due to forging action of shoulder that pushed the flowing material downward for its proper consolidation behind the tool pin (Figure 4.6c and f).

Partial recrystallization was observed in TMAZ, which is attributed to lower temperature and insufficient plastic deformation as compared to those required for complete recrystallization of deformed grains (Figure 4.6b and e). HAZ is characterized by the region that experienced only thermal cycle without undergoing plastic deformation and resulted in coarse grains (Figure 4.6a and d). Nonhomogeneous plastic deformation occurred in SZ due to difference in properties of parent materials. Steuwer et al. (2006) reported that uneven heat distribution affects the tensile properties of dissimilar FS welds of aluminum alloy. Clear interface of AA2219 and AA7475 is observed in dissimilar joint, which may be attributed to nonhomogeneous mixing of both the alloys (Figure 4.7). However, dissimilar joint is fractured from TMAZ/HAZ of AS where lowest microhardness is found. This shows that weld is defect-free and nonhomogeneous mixing of BM does not degrade the joint strength.

Hardness profile across the weld, as shown in Figure 4.8, indicates that lowest microhardness is found at TMAZ/HAZ of RS (softer material). Microhardness of FS-welded joints of heat-treatable aluminum alloys depends on grain size and its orientation, distribution of precipitates, and dislocation density. Dissolution and coarsening of strengthening precipitate and grain growth is responsible for lower hardness values in TMAZ and HAZ of the welded joint of heat-treatable aluminum alloys. AA7475 is precipitation hardenable aluminum alloy, and its mechanical properties are derived mainly from the strengthening precipitates present in the microstructure. However, mechanical properties of AA2219-O (annealed condition) mainly depend on its grain size and orientation rather than on precipitates. Improved hardness is obtained in SZ of AA2219 side due to grain refinement. Abrupt change in hardness observed in the SZ of dissimilar joint indicates the presence of a clear interface between hard and soft alloy at SZ.

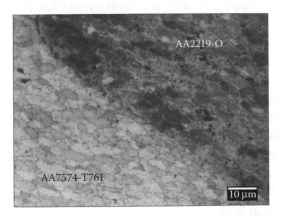

FIGURE 4.7   Microstructure of dissimilar FS-welded joint interface in SZ.

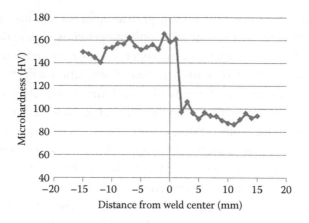

FIGURE 4.8   Microhardness distribution of AA7475-T761 and AA2219-O dissimilar weld.

4.8 FSW OF 6XXX–7XXX SERIES ALUMINUM

Several industrial sectors including railways, automotive, and aerospace industries use AA6xxx and AA7xxx Al alloys for making structural components. Dissimilar joining of these alloys is commonly required for fabrication of different parts of aircraft and space shuttles. FSW of dissimilar 6xxx–7xxx series aluminum alloys has been successfully used in various engineering applications with good joint properties. Dissimilar FSW of 6xxx–7xxx series aluminum alloys is presented in Table 4.3.

FSW of AA6061 and AA7075 was investigated and effect of process parameters and material position on microstructure and mechanical properties was analyzed (Guo et al., 2014). Figure 4.9 shows macrostructures of AA6061–AA7075 dissimilar joints welded under different welding conditions in which AA6061 appears darker and AA7075 appears brighter due to difference in etching responses. Effective material mixing is observed in the weld fabricated with AA6061 on AS (Figure 4.9b and d). AA7075, being harder material, is difficult to move properly from AS to RS due to its higher flow stress under given welding conditions. Similar results were obtained by Park et al. (2010) where they observed that during FSW of AA6061 and AA5051, materials were properly mixed by keeping AA5051 in AS. The position of material in dissimilar FSW of aluminum alloys affects the mechanical properties of welded joints. Some researchers suggested that higher strength materials should be located at AS to enhance the joint strength (Gerard and Ehrstrom, 2004; Amancio-Filho et al. 2008). Joint properties significantly depend on the material placed

TABLE 4.3 Dissimilar FSW of 6xxx–7xxx Series Aluminum Alloys

Material	Thickness (mm)	Rotation Speed (rpm)	Welding Speed (mm/min)	Pin Profile	References
6061-T6 and 7075-T6	6.3	1200	120, 180, 300.	Threaded conical, with three flats	Guo et al. (2014)
6061-T6 and 7075-T6 (both)	4.76	700, 850, 1000, 1150, 1300, 1450	100	Threaded and tapered with three flats	Cole et al. (2014)
7075-T6–6082-T6(AS)	8	1000, 1200	90, 120	Triangular frustum pin	Aval (2015)
6061-T6 7050-T7451(AS)	5	270, 340, 410	114	Threaded cylindrical	Rodriguez et al. (2015)
7075-T7351 (AS) and 6056	5	900	250	–	Srinivasan et al. (2005)

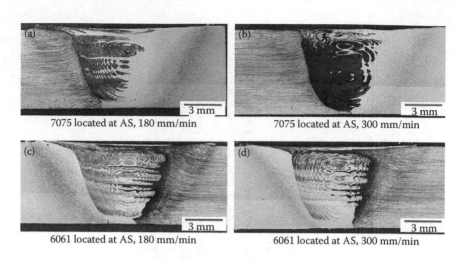

7075 located at AS, 180 mm/min	7075 located at AS, 300 mm/min
6061 located at AS, 180 mm/min	6061 located at AS, 300 mm/min

FIGURE 4.9 Macrostructure of the joints produced under different welding conditions. (Reprinted from *Materials and Design*, 56, J.F. Guo et al., Friction stir welding of dissimilar materials between AA6061 and AA7075 Al alloys effects of process parameters, 185–192, Copyright 2014, with permission from Elsevier.)

on RS because SZ is mainly composed of material placed at RS (Lee et al., 2003b). Welded joints clearly revealed onion ring structure under all welding conditions considered in the study performed by Guo et al. (2014). Typical multiple vortex centers were observed in the joints by employing threaded pin profile with three flats.

Microstructure of BMs and SZ of welded joints are shown in Figure 4.10. Fine and equiaxed grain in SZ of all joints showed that both alloys have undergone dynamic recrystallization. Equiaxed grains of less than 6 μm average diameter were observed in SZ toward AA7075 side, which was reduced to 3 μm by increasing welding speed from 180 to 300 mm/min. This may be due to lower peak temperature and shorter heat-holding time at increased welding speed that obstructed the grain growth (Figure 4.10b and c). Equiaxed grains of less than 10 μm average diameter were observed in SZ toward AA6061 side, which were reduced to 5 μm by increasing welding speed from 180 to 300 mm/min.

Refined grains in SZ toward AA7075 side are smaller than those of AA6061 side in same weld (Figure 4.10b and e). Difference in grain size between two alloys at SZ during FSW of AA7075 and AA6056 was also noticed by Srinivasan et al. (2007) and they revealed that this is due to initial grain size of parent material (AA7075). The presence of second-phase particles provides nucleation sites and restricts grain growth during

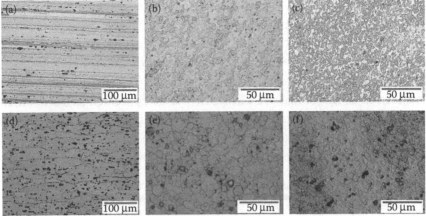

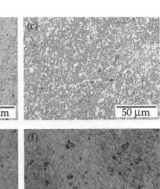

FIGURE 4.10   Microstructure of (a) base 7075, (b) SZ of 7075 side (180 mm/min), (c) SZ of 7075 (300 mm/min), (d) base 6061, (e) SZ of AA6061 (180 mm/min), and (f) SZ of 6061 (300 mm/min). (Reprinted from *Materials and Design*, 56, J.F. Guo et al., Friction stir welding of dissimilar materials between AA6061 and AA7075 Al alloys effects of process parameters, 185–192, Copyright 2014, with permission from Elsevier.)

recrystallization. The content of second-phase particle in AA7075 was higher compared to that of AA6061, which resulted in relatively lower grain size toward AA7075 side than AA6061.

FSW of age-hardened aluminum alloys results in reduction of microhardness in SZ compared to BM due to dissolution, coarsening, and reprecipitation of strengthening precipitates. However, grain refinement enhances the microhardness values but it does not significantly alter microhardness for age-hardened aluminum alloys. HAZ exhibits slightly lower hardness due to coarsening of strengthening precipitates and disappearance of G.P. zones and highest minimum value is obtained in the joints fabricated with least heat input condition. Lower hardness in TMAZ is observed because of severe coarsening and dissolution of precipitates. SZ experiences dissolution and coarsening of precipitates and also reprecipitation of some precipitates take place followed by dissolution due to highest peak temperature at SZ.

Tensile samples of dissimilar welded joint between AA7075 and AA6061 were fractured from HAZ on AA6061 side where minimum hardness existed. In dissimilar FSW of Al alloys, joints were usually fractured from HAZ on weaker material side (Amancio-Filho et al., 2008; Silva et al., 2011). Highest UTS was obtained at higher welding speed when AA6061 was located at AS probably due to less coarsening of precipitates in HAZ caused by less heat input (Guo et al., 2014).

FSW of 8-mm thick dissimilar aluminum alloys (AA7075-T6 and AA6082-T6) was performed at different rotational speed (1000 and 1200 rpm) and welding speed (90 and 120 mm/min) in which AA6082 was located at AS (Aval, 2015). Macrostructure of welded joints at different rotational and welding speeds is shown in Figure 4.11. Complex patterns were observed in the joint fabricated with higher rotational speed (1200 rpm) and lower welding speed (90 mm/min) due to intense material

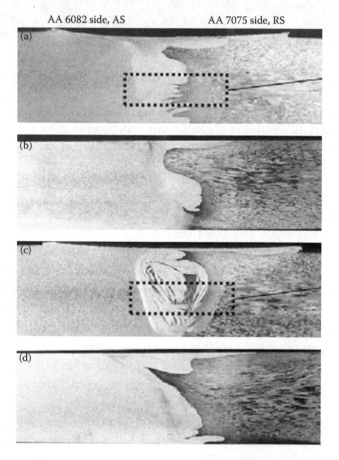

FIGURE 4.11 Macrostructure of FS-welded joints produced at different rotational and welding speeds of (a) 1000 rpm, 90 mm/min, (b) 1000 rpm, 120 mm/min, (c) 1200 rpm, 90 mm/min, and (d) 1200 rpm, 120 mm/min. (Reprinted from *Materials and Design*, 87, H.J. Aval, Microstructure and residual stress distributions in friction stir welding of dissimilar aluminum alloys, 405–413, Copyright 2015, with permission from Elsevier.)

mixing in SZ (Figure 4.11c). However, such patterns were not obtained at lower rotational speed. Joints fabricated with lower welding speeds showed better material mixing, which may be attributed to comparatively higher heat input in samples A and C than others.

Microstructure of HAZ was similar to that of BM in both sides. Bent and elongated grains observed in TMAZ were more visible in AS compared to that in RS due to large degree of relative deformation at AS (Aval, 2015). Intense plastic deformation and high heat input at SZ during FSW resulted in recrystallized fine grains and dissolution, coarsening and reprecipitation of strengthening precipitates. Smaller grain size at AA7075 side in SZ was observed compared to that at AA6082 side. Similar grain size difference in SZ during dissimilar FSW of 6000 and 7000 series alloys was also reported by Srinivasan et al. (2005). This difference in grain size at SZ was attributed to the initial grain size of BM and higher content of second-phase particles in AA7075. Degree of plastic deformation and peak temperature during FSW also affect the grain size of SZ. For instance, smaller grain size (4.9 μm) of SZ in the AA6082 side of joint produced with low heat input (1000 rpm, 120 mm/min) was obtained while joint produced with high heat input (1200 rpm, 90 mm/min) resulted in larger grain size (8.4 μm). Therefore, higher temperature at AS (AA6082 side), higher second-phase particles in AA7075, and finer grain size in parent AA7075 resulted in fine grains in AA7075 side of SZ compared to AA6082 side (Aval, 2015).

Comparison between microhardness values of joints produced with lowest heat input (sample B; 1000 rpm and 120 mm/min) and highest heat input (sample C; 1200 rpm and 90 mm/min) after natural aging of 365 days is shown in Figure 4.12. Softening took place in the joint produced with high heat input resulted in lower hardness values compared to that produced with low heat input. In sample C (Figure 4.11c), effective mixing of BMs is observed and AA6082 moved to AA7075 side due to stirring of the pin leading to lower hardness values in AA7075 side at 2 mm away from the weld center (Figure 4.12). TEM micrographs of BMs and welded zones of sample B are depicted in Figure 4.13. Strength of welded joint between 6xxx–7xxx series aluminum alloys is largely associated with precipitation of Guinier-Preston's (GP) zones, β'' and η' phases in SZ.

TEM micrograph of AA7075-T6 BM revealed fine distribution of precipitates of 10–90 nm in size (Figure 4.13). Finer precipitates (lower than 20 nm) are GP zones and coarser precipitates are η' (Rhodes et al. 1997; Su et al. 2003). TEM micrograph of AA6082-T6 BM revealed fine distribution

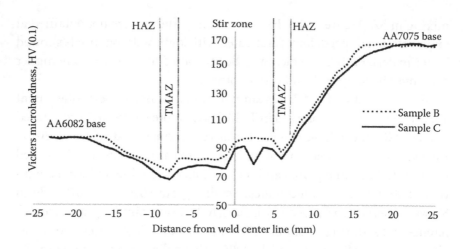

FIGURE 4.12 Comparison between microhardness values of joints produced with 1000 rpm and 120 mm/min (sample B) and 1200 rpm and 90 mm/min (sample C) after natural aging of 365 days. (Reprinted from *Materials and Design*, 87, H.J. Aval, Microstructure and residual stress distributions in friction stir welding of dissimilar aluminum alloys, 405–413, Copyright 2015, with permission from Elsevier.)

of precipitates of 10–50 nm in length. Morphology of these precipitates in Figure 4.13b indicates β'' phase (Olea et al., 2007).

Coarse round particles of average size 70–250 nm are observed in TMAZ of AA7075 side after dissolution of small precipitates (Figure 4.13c). Coarser and wider precipitates of round and rod morphology were observed in TMAZ of AA6082 side. Rod-shaped precipitates of 80–250 nm in length and 9–15 mm in width belong to β' phase and round precipitates of 60 nm average size belong to Si and β phases (Figure 4.13d). Reduction in microhardness values in TMAZ of the joint compared to BM was attributed to the change in type and morphology of TMAZ precipitates. Growth of BM precipitates (GP and η' in AA7075 and β'' in AA6082) occurred in TMAZ, which resulted in decrease in microhardness. Three different precipitates (GP zones, round type η', and rod-shaped η precipitates) were observed in SZ of AA7075 side and round-type β and Si and rod-type β' precipitates were observed in SZ of AA6082 side (Figure 4.13f). Higher microhardness values in SZ after natural aging were associated with reprecipitation of fine precipitates β' and η' and GP zones. Reprecipitation of β', GP, and η' occurred in the SZ, which resulted in increase in microhardness values as shown in Figures 4.12 and 4.13e, f.

Microstructure and mechanical properties of dissimilar FS-welded AA6061–AA7050 at different rotational speed (270, 340, and 410 rpm) were

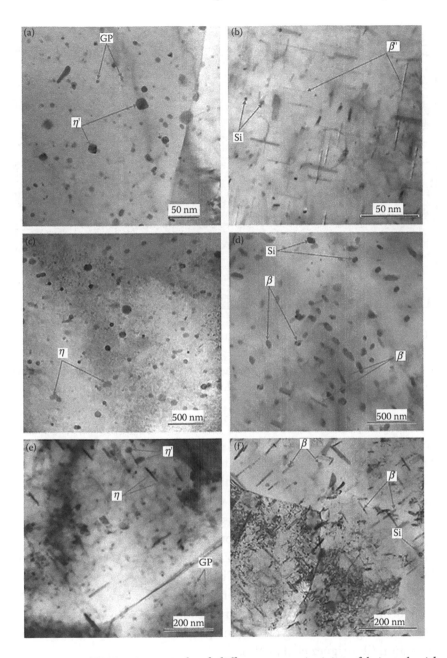

FIGURE 4.13 TEM micro graph of different zones in joints fabricated with 1000 rpm and 1200 mm/min; (a) AA7075-T6 BM, (b) AA6082-T6 BM, (c) TMAZ of AA7075 side, (d) TMAZ of AA6082 side, (e) SZ of AA7075 side, and (f) SZ of AA6082 side. (Reprinted from *Materials and Design*, 87, H.J. Aval, Microstructure and residual stress distributions in friction stir welding of dissimilar aluminum alloys, 405–413, Copyright 2015, with permission from Elsevier.)

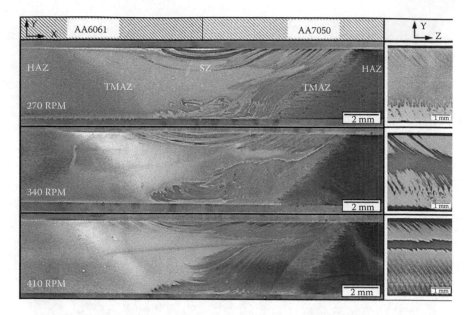

FIGURE 4.14 Macrostructure of dissimilar AA6061–AA7050 FS-welded joints at different tool rotational speeds. (Reprinted from *Materials and Design*, 83, R.I. Rodriguez et al., Microstructure and mechanical properties of dissimilar friction stir welding of 6061-to-7050 aluminum alloys, 60–65, Copyright 2015, with permission from Elsevier.)

investigated (Rodriguez et al., 2015). The transverse (X–Y) sections of the joints fabricated at different rotational speeds are shown in Figure 4.14. Both alloys are distinguishable because of difference in etching response. AA6061 and AA7050 are shown as bright and dark-colored regions, respectively.

In this case also, three characteristically different regions corresponding to SZ, TMAZ, and HAZ were formed. Vortex structure consisting of alternate lamella was observed in SZ due to intimate mixing of both the BMs by the threaded tool. Grains of both alloys were significantly refined in SZ due to SPD produced by stirring. Refined equiaxed grains in SZ were formed due to dynamic recrystallization. TMAZ was characterized by elongated grains having undergone plastic deformation but not recrystallization (Rodriguez et al., 2015).

Three distinct layers in SZ are observed in Figure 4.15. Compositions at location A and location B are found to be similar to that of AA6061 and AA7050, respectively. Location C consisted of a combination of both materials. Similar observations have been demonstrated by other researchers as well for AA6061–AA7075 (Guo et al., 2014).

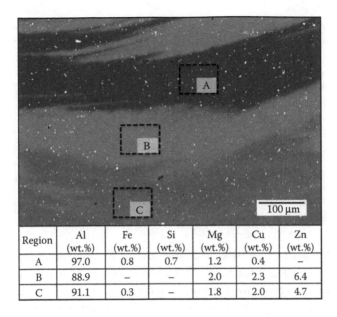

Region	Al (wt.%)	Fe (wt.%)	Si (wt.%)	Mg (wt.%)	Cu (wt.%)	Zn (wt.%)
A	97.0	0.8	0.7	1.2	0.4	–
B	88.9	–	–	2.0	2.3	6.4
C	91.1	0.3	–	1.8	2.0	4.7

FIGURE 4.15   SEM micrograph of three distinct layers observed in SZ of dissimilar weld. (Reprinted from *Materials and Design*, 83, R.I. Rodriguez et al., Microstructure and mechanical properties of dissimilar friction stir welding of 6061-to-7050 aluminum alloys, 60–65, Copyright 2015, with permission from Elsevier.)

Microstructure of longitudinal section of the joints revealed that the spacing between the material bands decreased with the increase in rotational speed. Band spacing of 460, 350, and 300 μm were obtained at 270, 340, and 410 rpm, respectively. Asymmetric hardness profile was obtained at all rotational speeds due to difference in mechanical properties of BMs. Strength and hardness of dissimilar joints were affected by dissolution of precipitates into the matrix at SZ and coarsening of precipitates at HAZ. Tensile strength increased with increase in the rotational speed and all the joints were fractured from AA6061 side (Figure 4.16). Other dissimilar FSW studies also reported similar results of joint failure at softer material side (Amancio-Filho et al., 2008; Silva et al., 2011). The joint produced with lower rotational speed (270 rpm) failed in the region adjacent to SZ of RS while joints produced with medium (340 rpm) and high (410 rpm) rotational speeds failed in TMAZ/HAZ boundary of RS. The joint produced with low rotational speed featured large voids located at bottom portion of weld due to inadequate mixing and transport of material during welding. Failure observed through HAZ in joint fabricated at high rotational speed was due to material softening and coarsening of strengthening precipitates.

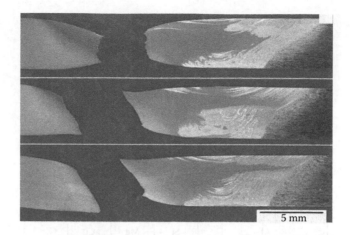

FIGURE 4.16 Macrostructure of fractured sample of joints produced at different rotational speeds. (Reprinted from *Materials and Design*, 83, R.I. Rodriguez et al., Microstructure and mechanical properties of dissimilar friction stir welding of 6061-to-7050 aluminum alloys, 60–65, Copyright 2015, with permission from Elsevier.)

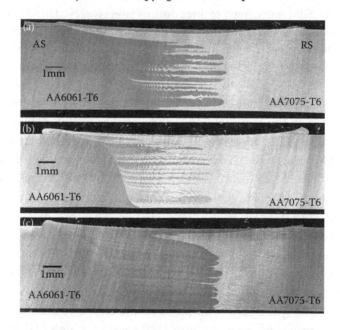

FIGURE 4.17 Macrostructure of dissimilar FSW of AA6061–AA7075 at tool off-set at (a) 0 mm, (b) 2 mm toward AA7075-T6, and (c) 2 mm toward AA6061-T6. (With kind permission from Springer Science+Business Media: Weld temperature effects during friction stir welding of dissimilar aluminum alloys 6061-T6 and 7075-T6, *International Journal of Advanced Manufacturing Technology*, 71, 2014, 643–652, E.G. Cole et al.)

Macrostructure of dissimilar (AA6061-T6 and AA7075-T6) FS-welded joints fabricated with different tool offset values is shown in Figure 4.17, which reveals that the amount of both the materials in SZ is significantly influenced by tool pin offset (Cole et al., 2014). The joint produced with tool offset toward stronger material resulted in higher strength. The amount of material (AA7075) swept from RS into SZ increased by increasing offset toward stronger material and accordingly joint strength increased and showed positive trend with amount of AA7075 in SZ (Cole et al., 2014). Nearly, all joints produced at different tool offset values fractured from HAZ of AS (AA6061).

Lower heat was generated at softer material (AA6061) side when tool offset was given toward stronger material (AA7075). This resulted in less coarsening of precipitates in HAZ of softer material, which in turn enhanced the joint strength. Strength of weaker material was preserved in dissimilar weld by giving the tool offset toward stronger material.

CHAPTER 5

Case Study on AA5083–AA6063 Dissimilar Welding

5.1 INTRODUCTION

Different components of a product perform varied functions and are always subjected to different service conditions. Fabrication of large and complex products requires assembly and joining of more than one component, often made of different materials. Thus, welding of dissimilar materials becomes essential to produce components that can withstand complex service conditions. Unfortunately, joining of dissimilar materials is difficult and marred with a high probability of imperfections, which makes it necessary to employ sound welding techniques for good results and also demands a thorough understanding of imperfections and associated causes so that they can be either eliminated or minimized. The issue of imperfections at joints becomes predominant when difference in properties of alloys, which is actually often too much when heat-treatable and non-heat-treatable aluminum alloys are welded, is large making mixing and joint consolidation difficult.

5.2 ISSUES IN DISSIMILAR MATERIALS JOINING BY FSW

Recently, adoption of FSW has seen enormous increase across different industrial sectors such as automobile, aerospace, shipbuilding, etc. and this has increased research interest in the process particularly with an aim to better understand FSW factors that lead to defects in joints (Nakata et al., 2000; Lee and Jung, 2003c; Khodir and Shibayanagi, 2008; Dehghani et al., 2013). Two common defects namely tunneling defect and KB are found in FSW joints and these defects are quite different from those found in fusion welding. Improper selection of typical FSW parameters such as tool design, tool rotation and traverse speed, depth of tool plunge, angle of tool tilt, tool pin offset, joint gap, etc. may lead to defect formation. A reasonable volume of literature is available that suggests that pin offset plays a vital role in affecting weld quality of a dissimilar FSW joint (Lee and Jung, 2003c; Dehghani et al., 2013; Cole et al., 2014) and hence it needs to be carefully selected taking into consideration physical, mechanical, metallurgical, and thermal properties of the materials being joined. In addition, the amount of plunge depth is also carefully chosen during FSW of dissimilar materials as it plays a significantly role during plastic deformation at the time of stirring.

Interestingly, most of the researches considered only conventional FSW process parameters and investigated their effects on defect formation during FSW of dissimilar materials. It may be noted that there is significant difference in several properties of the dissimilar materials being joined, such as softening temperature, strength, thermal conductivity, heat capacity, chemistry, and metallurgy. Greater difference in these properties of dissimilar materials constitutes a major source of joint imperfections. Further, during welding with conventional FSW parameters, the conditions of heat and material flow (stirring) will be the same for the two materials but there will be difference of softening and flow stress. Strategically, different magnitude and position of offset during FSW of dissimilar materials should be chosen to produce tailor-made desired unequal effects on both side of the material.

5.3 TYPICAL PAIR OF DISSIMILAR MATERIALS

A typical pair of AA5083 (non-heat-treatable) and AA6063 (heat-treatable) despite both being alloys of aluminum, have large difference in mechanical properties as illustrated in Table 5.1. The AA5083 is one of the strongest non-heat-treatable aluminum alloys and is used in shipbuilding (due to corrosion resistance to seawater and chemical attacks), rail cars, pressure vessels, vehicle bodies, mine skips, cages, etc. The AA6063 is a heat-treatable alloy and finds extensive application in road and rail transport, extreme sport equipment, architectural applications, hydro-formed tube for chassis, etc.

TABLE 5.1   Mechanical and Thermal Properties of AA5083-H116 and AA6063-T6

Aluminum Alloy	UTS (MPa)	Yield Strength (MPa)	Elongation (%)	Thermal Conductivity (W/mK)	Melting Point (°C)	Fatigue Strength
AA5083	350	145	23	120	591	150
AA6063	220	110	14	200	616	70

In such a situation of great difference in properties, the behavior of each material during welding will be grossly different. This is because, the FSW tool will produce a common condition of heat and material flow speed for both materials, yet each material's response in terms of softening temperature and flow stress will be highly different. Consequently, material consolidation at joint will be affected resulting in a defective joint.

5.4 CASE STUDY OF AA5083 AND AA6063 DISSIMILAR WELDING

Studies that are focused on investigation of defect formation may provide solution, which may prevent imperfections in most dissimilar joint. In line with this, a case study is presented here that analyzes defects that may typically be introduced during FSW of dissimilar aluminum alloys AA5083 and AA6063. Apart from the application mentioned above, AA5083 finds large-scale application in fabrication of ship hulls owing to its good corrosion resistance to seawater and, AA6063 is widely used in making internal structure of ship as well as in automobiles because of its properties such as good workability, good corrosion resistance, light weight, and moderate strength. Taken together, these alloys exhibit suitable properties that have led to enormous welding activities between these materials in automobile and shipbuilding and, therefore, this specific pair of alloys has been chosen.

Generally, FSW process parameters such as tool pin offset and plunge depth are kept constant during FSW, although they significantly affect the weld quality of dissimilar joint. It is thus important to decide correct values of tool offset and plunge depth. These two parameters were varied with an aim to investigate their effect on formation of tunneling defect and KB so as to understand their contribution toward the formation of these defects. In addition, critical values of these parameters that yield minimum defects were also determined and dissimilar FSW was performed at these critical values along with optimum values of conventional parameters in order to produce good-quality joints. This case study would provide knowledge to readers about the effect of FSW process parameters on formation of tunneling and KB defects so that they can understand the possible reasons and take necessary corrective measures.

5.4.1 Experimentation Performed in the Investigation

For the investigation, 4.75-mm-thick plates of AA5083-H116 and AA6063-T6 were used as BMs. Table 5.1 lists the mechanical and thermal properties of these alloys. Rectangular weld samples of the alloys having 200 mm length and 45 mm width were made and AA5083 was kept in the AS and AA6063 on the RS.

A tungsten carbide tool with shoulder diameter of 20 and 4.40 mm long tapered conical pin was employed during welding, and rotational speed, traverse speed, and tool tilt angle were kept at 450 rpm, 100 mm/min, and 2°, respectively. The tool offset (A) and plunge depth (B) were varied during experimentation. Table 5.2 shows FSW experimental plan used in the present study.

Prior to welding, the prepared weld samples were thoroughly cleaned by organic solvent, washed and dried, and subsequently they were longitudinally butt welded on a retrofitted robust vertical milling machine (make: Bharat Fritz Werner, India) capable of performing FSW, as shown in Figure 5.1. Continuous variation in the tool offset was achieved by employing a special

TABLE 5.2 The FSW Experimental Plan

Weld Number	Pin Offset (A) (mm)			Plunge Depth (B) (mm)
	Toward AA5083	Zero Pin Offset	Toward AA6063	
S41	1.5			0.3
S42	1.0			0.3
S43	0.5			0.3
S44		0.0		0.3
S45			0.5	0.3
S46			1.0	0.3
S47			1.5	0.3
S31	1.5			0.4
S32	1.0			0.4
S33	0.5			0.4
S34		0.0		0.4
S35			0.5	0.4
S36			1.0	0.4
S37			1.5	0.4

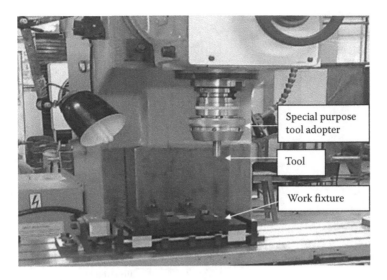

FIGURE 5.1 Experimental setup of FSW. (Reprinted from *Journal of Alloys and Compounds*, 648, N.Z. Khan et al., Investigations on tunneling and kissing bond defects in FSW joints for dissimilar aluminum alloys, 360–367, Copyright 2015b, with permission from Elsevier.)

purpose work fixture, which was indigenously designed and developed. The schematic of continuous variation of the tool offset is shown in Figure 5.2. The tool offset was varied between 0 and 1.5 mm on both sides (i.e., AA5083 side and AA6063 side) and plunge depth was varied from 0.3 to 0.4 mm.

5.4.2 Analysis of Defect Formation

Defects and their effects can be characterized and analyzed through a comprehensive regime of mechanical testing and microstructural analysis. Microstructural characterization of the area around defects helps in defining size, location, and morphology of the defect. This macro/microstructural characterization, when correlated to the mechanical characterization leads us in understanding the effects of defects on joint quality.

5.4.2.1 Tunneling Defect

A common defect associated with FSW is tunneling defect, which is also referred to as wormhole defects owing to its similarity with a wormhole (Figure 5.3). Due to this defect, mechanical properties of joints are

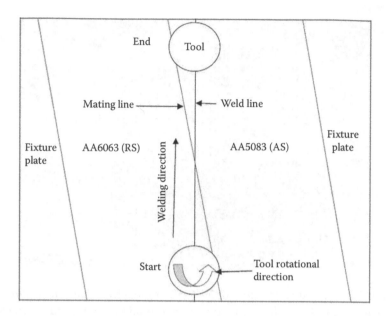

FIGURE 5.2   Schematic diagram showing continuous variation of tool pin offset during FSW. (Reprinted from *Journal of Alloys and Compounds*, 648, N.Z. Khan et al., Investigations on tunneling and kissing bond defects in FSW joints for dissimilar aluminum alloys, 360–367, Copyright 2015b, with permission from Elsevier.)

FIGURE 5.3   Tunneling defect.

significantly affected. This defect is formed below weld face (exposed surface) and normally not visible. Like most other defects in FSW, wormhole is also formed due to improper selection of welding process parameters (e.g., welding rotation and traverse speed, applied pressure, tool offset, improper design of tool, etc.) (Zettler et al., 2010). A low heat input during FSW is also one of the causes for the formation of tunneling defects, as it results in high flow stresses causing material movement deficient which leads to defect formation (Kim et al., 2006). In order to either minimize or eliminate tunneling defect, there should be sufficient heat input and it should be appropriately distributed on respective sides.

The effect of tool pin offset on defect formation in the present case study was investigated by continuously varying it on both sides of joint, as shown in Figure 5.2. Welded joints were examined at various offset locations and tunneling defect was found at several locations. In all cases, defect was present in between weld nugget and TMAZ and it spread over AA5083 (AS). A close examination of samples at different offset locations (Figure 5.4) reveals that (i) welding done with improperly selected pin offset results in insufficient mixing leading to development of unequal flow stress in the two alloys and (ii) insufficient consolidation of flowing material behind

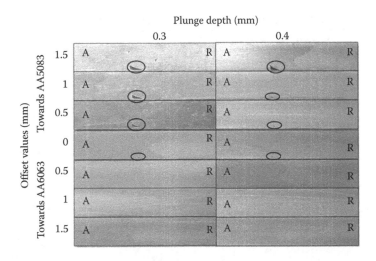

FIGURE 5.4 Tunneling defect at different values of tool pin offset and plunge depth. (Reprinted from *Journal of Alloys and Compounds*, 648, N.Z. Khan et al., Investigations on tunneling and kissing bond defects in FSW joints for dissimilar aluminum alloys, 360–367, Copyright 2015b, with permission from Elsevier.)

the pin takes place due to inadequate plunge depth resulting in tunneling defect.

The tunneling defect cross section is approximately wedge shaped (Figure 5.3). During welding, maximum temperature reduces due to insufficient forging pressure, which in turn increases flow stress of both the materials leading to insufficient plastic deformation and a sluggish material movement. The main reason for the formation of the tunneling defect is improper mixing and inadequate consolidation of material at joint, which happens when FSW is performed with zero pin offset and also with offset given toward stronger material (AS). Further, the forward movement of tool before depositing enough material behind may create void, which indeed is manifested as tunneling defect. Mechanical properties of the joint get degraded due to increasing size of tunneling defect.

A close examination of Figure 5.4 and Table 5.3 shows that joint strength is inversely proportional to size of tunneling defect. The results in the present case demonstrate that the joints fabricated with offset toward weaker material (RS) are free from tunneling defects. It is also observed that the size of tunneling defect is correlated with plunge depth as well, such that an increase in plunge depth from 0.3 to 0.4 mm caused reduction in size of the

TABLE 5.3   Tensile Properties and Fracture Locations of the Welded Joints

Weld Number	Properties at Room Temperature		Fracture Location	Specimen Images
	Tensile Strength (MPa)	Elongation (%)		
S41	65.8	12	SZ	
S42	44.8	4.8	Interface	
S43	85	7.8	Interface	
S44	71	12.25	Interface	
S45	137	29	SZ	
S46	136.8	11.25	TMAZ of AA5083	
S47	121	9.1	TMAZ of AA5083	
S31	28.5	11.08	TMAZ of AA6063	
S32	28.9	5.58	TMAZ of AA6063	
S33	98.4	8	SZ	
S34	156.1	5.58	Interface	
S35	172.5	9.4	SZ	
S36	145.9	14.7	TMAZ of AA5083	
S37	103.3	4.75	TMAZ of AA5083	

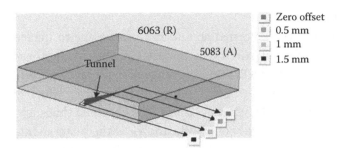

FIGURE 5.5   3D model reconstructed by stitching various tunnel sections at 0.4 mm plunge depth. (Reprinted from *Journal of Alloys and Compounds*, 648, N.Z. Khan et al., Investigations on tunneling and kissing bond defects in FSW joints for dissimilar aluminum alloys, 360–367, Copyright 2015b, with permission from Elsevier.)

defect. Figures 5.5 and 5.6 are the three-dimensional (3D) views prepared using CAD software, which exhibit effect of continuously increasing offset on size of tunneling defects for two different values of plunge depth.

It can be seen from Figures 5.5 and 5.6 that an increase in plunge depth as well as offset values toward weaker material results in reduction in size of tunneling defect. This happens because of associated increase in heat input, adequate forging pressure and proper mixing. Combined effect of plunge depth and tool pin offset can provide adequate forging pressure and proper mixing, which may reduce tunneling defect and enhance the joint strength.

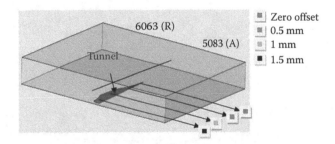

FIGURE 5.6   3D model reconstructed by stitching various tunnel sections at 0.3 mm plunge depth. (Reprinted from *Journal of Alloys and Compounds*, 648, N.Z. Khan et al., Investigations on tunneling and kissing bond defects in FSW joints for dissimilar aluminum alloys, 360–367, Copyright 2015b, with permission from Elsevier.)

5.4.2.2 KB Defect

Inefficient material deformation and movement due to inadequate pressure and insufficient stirring of material results in another solid state bonding defect called KB. In the presence of oxide layer on BM, the conditions mentioned above will lead to improper bonding, which gives rise to the formation of KB (Chen et al., 2006). It is observed that such a discontinuity exists at either the root of weld and/or at the material interface.

It is evident from Figure 5.7 that either a little or no metallic bonding between two previously separated regions of BMs exists in KB. Figure 5.7b–d shows magnified SEM micrograph of sample S34, which depicts the presence of tunneling defect and KB. Aluminum alloys have high affinity to oxygen and, therefore, during FSW it is extremely important to remove oxide interface between abutting plates so that an intimate bond without KB is formed. The problem becomes more prominent if tool pin does not penetrate the entire thickness of the BM plates (Oosterkamp et al., 2004; Chen et al., 2006). KB defect adversely affects the mechanical properties of joint, as it is evident from Table 5.3, which reveals that in samples fabricated with offset toward weaker material tunneling defect is absent but KB is prominent due to which their strength is poor.

Figure 5.8 shows the UTS of all weld samples. It is evident from this figure that UTS of joints fabricated at offset values toward stronger material

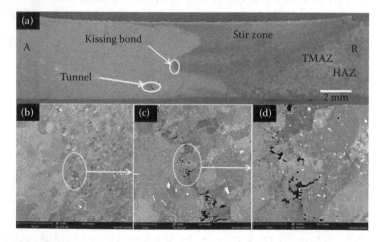

FIGURE 5.7 SEM micrograph showing KB. (Reprinted from *Journal of Alloys and Compounds*, 648, N.Z. Khan et al., Investigations on tunneling and kissing bond defects in FSW joints for dissimilar aluminum alloys, 360–367, Copyright 2015b, with permission from Elsevier.)

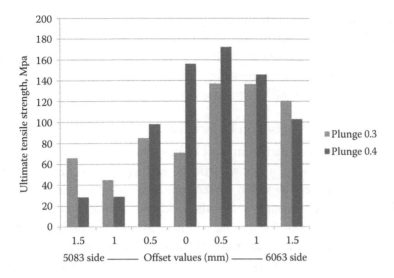

FIGURE 5.8   UTS (in MPa) of joints at different parameter combinations. (Reprinted from *Journal of Alloys and Compounds*, 648, N.Z. Khan et al., Investigations on tunneling and kissing bond defects in FSW joints for dissimilar aluminum alloys, 360–367, Copyright 2015b, with permission from Elsevier.)

is poor and it gets further adversely affected due to the presence of both the defects.

5.4.3   Important Parameters and Their Effects on Material Being Consolidated

The two important FSW parameters that varied in this case study play different roles. The amount of heat generated during FSW, forging action of the tool, and welding thrust are controlled by plunge depth, whereas distribution of generated heat and mixing of two materials are significantly affected by the tool offset. Dynamic recrystallization consequent to SPD during stirring action improves the joint strength and toughness simultaneously. An increase in tool rotation and plunge depth, and decrease in traverse speed result in increase in heat input. But, excessive high heat input condition favors the formation of IMCs and supports grain growth (Dickerson et al., 2003). IMCs being hard and brittle reduce the strength and ductility of joint. In case of lower heat input, softening of BMs gets adversely affected, which lead to inefficient consolidation and inadequate mixing. Thus, there has to be an appropriate amount of heat input in order to produce good-quality joints by FSW. Increasing the plunge depth alone to prevent defect formation does not seem to be a correct strategy. Rather, variation of plunge depth and tool pin offset while keeping the tool rotation and traverse speed constant would increase associated heat input and also achieve adequate forging action and flow stress balance. This strategy may be of great help in suggesting a choice of offset and plunge depth to either minimize or totally eliminate tunneling and KB defects.

5.4.4 Effect of Pin Offset on Defect Formation

It is pertinent that different amount of heat is required to soften different materials. It may happen that a particular setting of FSW parameters produces sufficient heat to soften weaker materials but the heat may not be sufficient to soften stronger materials. Evidently, a chosen setting of FSW parameters in a dissimilar joining will have same effect on both materials that behave differently. An appropriate amount of offset given toward weaker material may move relatively lesser amount of stronger material giving rise to effective stirring of both materials. Thus, strategic selection of an appropriate pin offset toward weaker material is important for dissimilar welding.

5.4.4.1 Effect of Pin Offset on Tunneling Defect

Poor-quality joints with defects have been reported by researchers when dissimilar FSW was performed with zero offset (tool axis on the joint line) (Murr et al., 1998; Ouyang et al., 2006). Dissimilar material welding is complicated because difference in mechanical, physical, and thermal properties does not allow equal softening of the two materials at same heat input. The dissimilar materials used in the present case possess different mechanical and thermal properties. In this situation, process parameter with zero pin offset generates equal heat in both materials leading to unequal softening and an imbalance in flow stress. For successful joining of dissimilar materials, different heat distribution is required for balancing flow stresses. Consequently, offset value is chosen in such a way that the tool stirs relatively more amount of weaker material (which is softer as well) to enable adequate mixing of two materials during stirring. Establishing heat balance between weaker and stronger materials during stirring and transport of materials results in balancing of flow stresses, which in turn will not only prevent formation of tunneling defect but also will improve the joint strength. The presence of tunnel in joint reduces load-carrying cross-sectional area and acts as a source for stress concentration leading to reduction in mechanical properties and fracture of the joint starting from this location. Figure 5.5 also reveals the formation of tunneling defect in all joints except where tool offset is given toward weaker material.

(a) (b)

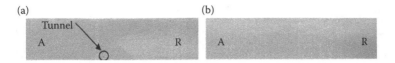

FIGURE 5.9   Macrograph depicting transverse section of the weld at pin offset (a) 0.0 (with tunnel, for S34) and (b) 0.5 mm (without tunnel for S35). (Reprinted from *Journal of Alloys and Compounds*, 648, N.Z. Khan et al., Investigations on tunneling and kissing bond defects in FSW joints for dissimilar aluminum alloys, 360–367, Copyright 2015b, with permission from Elsevier.)

Figure 5.9a and b presents a comparison between joints with defects and the one without it. Figure 5.9a shows a welded joint that was obtained at zero offset, whereas Figure 5.9b depicts the joint that was obtained at 0.5 offset toward weaker material. These figures show that tunnel is present in joint welded with zero offset and it was not formed when the offset was ≥ 0.5 mm. This is simply because when larger offset is given toward AA6063, then tool stirs small amount of AA5083, which is a relatively stronger material, and it becomes easy for the tool to mix small amount of AA5083 with the AA6063 in greater proportion in the nugget zone. As a result, more stirred material gets moved and deposited behind the tool, which eliminates chances of tunnel formation. Thus, by increasing the offset toward weaker material, good consolidation and coalescence of the joint is obtained. Contrary to this, at zero offset an increased amount of AA5083 is stirred in nugget zone. Since at same temperature, AA5083 alloy is less soft as compared to AA6063 and flow stress for AA5083 is large, it is difficult to move larger amount of AA5083 and at the same time flow stress is also large. Consequently, flow, movement, and mixing of material become difficult leading to deposition of lesser amount of stirred material behind the tool, which causes formation of the tunnel. Thus, tunneling defect can be eliminated by offsetting the tool toward weaker material, which is also evident from Figure 5.5.

5.4.4.2 Effect of Pin Offset on KB Defect

This defect can be eliminated by thorough mixing of materials during welding such that interface oxide layer is whipped effectively. In dissimilar joining, proper mixing may not be possible at zero pin offset due to vast difference in material properties, flow stress requirement, and material movement and therefore, adequate pin offset is extremely important. The presence of KB severely affects the mechanical properties of joint. Figure 5.10a shows a Charpy test sample S31 in which both tunneling

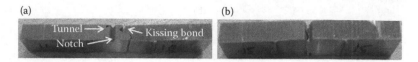

FIGURE 5.10   Fractured location of the Charpy impact test samples that were fractured at (a) interface (S31) and (b) notch (S35). (Reprinted from *Journal of Alloys and Compounds*, 648, N.Z. Khan et al., Investigations on tunneling and kissing bond defects in FSW joints for dissimilar aluminum alloys, 360–367, Copyright 2015b, with permission from Elsevier.)

defect and KB were present. It can be seen from this figure that the sample failed at KB rather than at notch or tunnel. This suggests that in comparison to tunneling defect, KB affects UTS and impact strength more severely. Figure 5.10b shows another Charpy test sample S35, which is a defect-free sample. It is evident from this figure that the sample fractured at notch suggesting good joint strength. Proper mixing between materials in S35 would have intimately whipped the oxide layer at an offset of 0.5 mm toward weaker material (KB absent), whereas improper mixing between materials in S31 would have resulted in the formation of KB at zero offset. Tensile test results also confirm that the presence of KB might be the reason for lower tensile strength of sample S31 (28.5 MPa), whereas tensile strength of sample S35 was found to be the greatest (at 172.5 MPa).

Foregoing analysis reveals that variation in the FSW process parameters, as done in the present investigation, has a major effect on the important phenomenon of material transport, heat distribution, and flow stress balance, which in turn affects the defect formation.

5.4.5 Effect of Plunge Depth on the Defect Formation

A lower plunge depth causes inadequate material flow due to less forging force and also bonding between dissimilar BM is poor, which leads to the formation of tunneling defect and KB adversely affecting the joint strength. However, an excessively high plunge depth results in excessive flash formation, joint thinning, and overheating. Overheating causes IMC formation, dissolution of strengthening precipitate, and grain coarsening causing reduction in strength. Therefore, it is necessary to use optimal value of plunge depth during FSW to obtain defect-free joint.

The material forging action and associated axial thrust in nugget is significantly affected by plunge depth. One study reports that defect-free joints during FSW of aluminum alloy AA7020-T6 can be produced by continuously increasing axial load beyond 8.1 kN (Kumar and Kailas, 2008a).

(a)   (b)

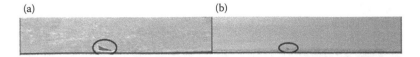

FIGURE 5.11   Effect of plunge depth on tunnel for sample (a) S32 and (b) S42. (Reprinted from *Journal of Alloys and Compounds*, 648, N.Z. Khan et al., Investigations on tunneling and kissing bond defects in FSW joints for dissimilar aluminum alloys, 360–367, Copyright 2015b, with permission from Elsevier.)

The present study was performed in velocity/position control mode and the thrust force was not measured, however, it was indirectly changed by varying the plunge depth.

The effect of plunge depth on size of tunnel was investigated by 14 experimental runs performed in the present case study and it was found that tunnel size is significantly affected by plunge depth. Further, to demonstrate the effect of plunge depth on tunnel, two samples, that is, S42 and S32 were welded using same set of FSW parameters except plunge depth of 0.3 and 0.4 mm, respectively, and the comparison is presented in Figure 5.11, which reveals that as plunge depth increases from 0.3 to 0.4 mm tunnel size reduces significantly. It has been observed in the present study that even a small decrease in the plunge depth by 0.1 mm (i.e., from 0.4 to 0.3 mm) gives rise to formation of defects, which in turn significantly reduces mechanical properties of the joint (S34—156.1 MPa to S44—71 MPa).

5.4.6 Effect of Plunge Depth and Pin Offset on Tensile Strength

Figure 5.8 shows tensile strength of the joints produced at different values of pin offset and plunge depth. Highest tensile strength is obtained for joint fabricated using 0.5 mm pin offset toward RS (weaker material). At fixed offset of 0.5 mm toward AA6063, joint fabricated using 0.4 mm plunge depth shows larger tensile strength compared to the one obtained by using 0.3 mm plunge depth. The main reason for reduced tensile strength is the tunneling defect as it is also evident from results of one study in which the authors observed significant increase in joint strength after the joint was repaired to correct the tunneling defect (Hui-jie and Hui-jie, 2009). Table 5.3 shows the tensile properties and fracture locations of welded joints, which reveal significant improvement in mechanical properties of the joint by using appropriate offset to Al 6063 (weaker material) side.

Figure 5.12 shows typical shape of fractured surface of samples that were fabricated at two different offset values. It is evident from Figure 5.12a that

(a)   (b)

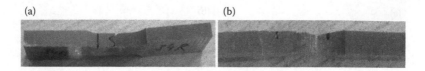

FIGURE 5.12   Fracture location of samples welded at (a) zero pin offset (S34) and (b) pin offset at 0.5 mm (S35). (Reprinted from *Journal of Alloys and Compounds*, 648, N.Z. Khan et al., Investigations on tunneling and kissing bond defects in FSW joints for dissimilar aluminum alloys, 360–367, Copyright 2015b, with permission from Elsevier.)

when offset is zero, the joint failed along KB present at interface in nugget. On the other hand, shape of fractured surface of a joint fabricated using offset of 0.5 mm toward AA6063 (defects-free joint) resembles a characteristic tensile failure, as revealed by Figure 5.12b.

5.5 SUMMARY OF THE CASE STUDY

Experimental investigation pertaining to defect formation during FSW of dissimilar aluminum alloys (AA5083-H116 and AA6063-T6) is presented in this study. Plates of AA5083-H116 and AA6063-T6 with 4.75 mm thickness were welded using FSW by employing a tapered cylindrical tool pin and effect of tool pin offset and plunge depth on the formation of tunneling defect and KB was investigated. Based on the results of the present study, the following conclusions are drawn:

- In FSW of dissimilar alloys tool pin offset and plunge depth play a vital role in the formation of defects (tunneling and KB defects) and therefore, they should be carefully selected.

- Insufficient mixing between AA5083-H116 and AA6063-T6 resulting from giving pin offset toward stronger material (AS) and also with zero offset leads to the formation of tunneling and KB defects.

- During FSW, heat generation is directly related to the plunge depth and by increasing the plunge depth heat generation is increased, which allows adequate mixing of dissimilar materials leading to improvement in the tensile strength consequent to minimization of defect formation.

- Mechanical properties of the welded joints are significantly affected by the KB defect as compared to that of tunneling defect.

- There is inverse relation between size of tunnels and plunge depth and therefore, tunnel size can be minimized by increasing plunge depth.

- Formation of tunneling defect can be prevented by offsetting the tool pin toward weaker material.

- Tunneling and KB defects affect the UTS of the joint when the joint is fabricated by offsetting the tool pin toward stronger material. However, only KB defect affects the UTS in case offset is given toward weaker material.

References

H. Abd El-Hafez. Mechanical properties and welding power of friction stirred AA2024-T35 joints. *Journal of Materials Engineering and Performance.* 20:839–845, 2011.

E.T. Akinlabi, A. Andrews, S.A. Akinlabi. Effects of processing parameters on corrosion properties of dissimilar friction stir welds of aluminium and copper. *Transactions of Nonferrous Metals Society of China.* 24(5):1323–1330, 2014.

S.T. Amancio-Filho, S. Sheikhi, J.F.D. Santos, C. Bolfarini. Preliminary study on the microstructure and mechanical properties of dissimilar friction stir welds in aircraft aluminium alloys 2024-T351 and 6056-T4. *Journal of Materials Processing Technology.* 206:132–142, 2008.

A. Amini, P. Asadi, P. Zolghadr. Friction stir welding applications in industry, in: M.K.B. Givi and P. Asadi (eds), *Advances in Friction Stir Welding and Processing.* Woodhead Publishing Limited, Cambridge, UK, 671–722, 2014.

W.J. Arbegast. Modeling friction stir joining as a metal working process, Presented at *Hot Deformation of Aluminum Alloys III*, San Diego, CA, USA, 2003.

M.M. Attallah, C.L. Davies, M. Strangwood. Influence of base metal microstructure on microstructural development in aluminium based alloy friction stir welds. *Science and Technology of Welding and Joining.* 12:361–369, 2007.

H.J. Aval. Microstructure and residual stress distributions in friction stir welding of dissimilar aluminium alloys. *Materials and Design.* 87:405–413, 2015.

H.J. Aval, S. Serajzadeh, A.H. Kokabi. Thermo-mechanical and microstructural issues in dissimilar friction stir welding of AA5086–AA6061. *Journal of Materials Science.* 46:3258–3268, 2011.

P. Bahemmat, M. Haghpanahi, M.K.B. Givi, K.R. Seighalani. Study on dissimilar friction stir butt welding of AA7075-O and AA2024-T4 considering the manufacturing limitation. *International Journal of Advanced Manufacturing Technology.* 59:939–953, 2012.

T.A. Barnes, I.R. Pashby. Joining techniques for aluminium space frames used in automobiles: Part I—Solid and liquid phase welding. *Journal of Materials Processing Technology.* 99:62–71, 2000.

J. Bevington. Spinning tubes mode of welding the ends of wire, rods, etc, and mode of making tubes. US Patent 463134, 1891.

H. Bouzaiene, M. Rezgui, M. Ayadi, A. Zghal. Correlation between welding and hardening parameters of friction stir welded joints of 2017 aluminum alloy. *Transactions of Nonferrous Metals Society of China.* 22:1064–1072, 2012.

M. Cabibbo, H.J. McQueen, E. Evangelista, S. Spigarelli, M. Di Paola, A. Falchero. Microstructure and mechanical property studies of AA6056 friction stir welded plate. *Materials Science and Engineering A.* 460–461:86–94, 2007.

B. Cai, D. Huang, B.L. Adams, T.W. Nelson. Microstructure characteristics of friction stir processes Al 7075 via macroscopic approach, in: *Frict Stir Weld Proc III TMS*, San Francisco, California, 269–275, 2005.

X. Cao, M. Jahazi. Effect of tool rotational speed and probe length on lap joint quality of friction stir welded magnesium alloy. *Materials and Design.* 32(1):1–11, 2011.

P. Cavaliere, A. De Santis, F. Panella, A. Squillace. Effect of welding parameters on mechanical and microstructural properties of dissimilar AA6082–AA2024 joints produced by friction stir welding. *Materials and Design.* 30(3):609–616, 2009.

P. Cavaliere, R. Nobile, F.W. Panella, A. Squillace. Mechanical and microstructural behaviour of 2024–7075 aluminium alloy sheets joined by friction stir welding. *International Journal of Machine Tools & Manufacture.* 46:588–594, 2006.

P. Cavaliere, F. Panella. Effect of tool position on the fatigue properties of dissimilar 2024–7075 sheets joined by friction stir welding. *Journal of Materials Processing Technology.* 206:249–255, 2008.

P. Cavaliere, A. Squillace, F. Panella. Effect of welding parameters on mechanical and microstructural properties of AA6082 joints produced by friction stir welding. *Journal of Materials Processing Technology.* 200:364–372, 2008.

D.J. Chakraborti, D.E. Laughlin. Phase relations and precipitation in Al–Mg–Si alloys with Cu additions. *Progress in Materials Science.* 49:389–410, 2004.

H.B. Chen, J.F. Wang, G.D. Zhen, S.B. Chen, T. Lin. Effects of initial oxide on microstructural and mechanical properties of friction stir welded AA2219 alloy. *Materials and Design.* 86:49–54, 2015.

H.B. Chen, K. Yan, T. Lin, S.B. Chen, C.Y. Jiang, Y. Zhao. The investigation of typical welding defects for 5456 aluminum alloy friction stir welds. *Materials Science and Engineering A.* 433:64–69, 2006.

S. Chen, X. Jiang. Texture evolution and deformation mechanism in friction stir welding of 2219Al. *Materials Science and Engineering A.* 612:267–277, 2014.

Z.W. Chen, T. Pasang, Y. Qi. Shear flow and formation of Nugget zone during friction stir welding of aluminium alloy 5083-O. *Materials Science and Engineering A.* 474:312–316, 2008.

E.G. Cole, A. Fehrenbacher, N.A. Duffie, M.R. Zinn, F.E. Pfefferkorn, N.J. Ferrier. Weld temperature effects during friction stir welding of dissimilar aluminum alloys 6061-T6 and 7075-T6. *International Journal of Advanced Manufacturing Technology.* 71:643–652, 2014.

J. Colligan, J. Paul Konkol, J. James Fisher, R. Pickens Joseph. Friction stir welding demonstrated for combat vehicle construction. *Welding Journal.* 82(3):34–40, 2003.

R. Crawford, G.E. Cook, A.M. Strauss, D.A. Hartman, M.A. Stremler. Experimental defect analysis and force prediction simulation of high weld pitch friction stir welding. *Science and Technology of Welding and Joining.* 11:657–665, 2006.

C.J. Dawes, W.M. Thomas. Friction stir process welds aluminum alloys. *Welding Journal.* 75:41–45, 1996.

T. DebRoy, H.K.D.H. Bhadeshia. Friction stir welding of dissimilar alloys: A perspective. *Science and Technology of Welding and Joining.* 15(4):266–270, 2010.

K. Dehghani, R. Ghorbani, A.R. Soltanipoor. Microstructural evolution and mechanical properties during the friction stir welding of 7075-O aluminum alloy. *International Journal of Advanced Manufacturing Technology.* 77:1671–1679, 2015.

M. Dehghani, A. Amadeh, S.A.A. Akbari Mousavi. Investigations on the effects of friction stir welding parameters on intermetallic and defect formation in joining aluminum alloy to mild steel. *Materials and Design.* 49:433–441, 2013.

S.S. Di, X.Q. Yang, G.H. Luan, B. Jiang. Comparative study on fatigue properties between AA2024-T4 friction stir welds and base materials. *Materials Science and Engineering A.* 435–436:389–395, 2006.

T.L. Dickerson, H.R. Shercliff, H. Schmidt. A weld marker technique for flow visualisation in friction stir welding. In: *Proc 4th Int Symp on Friction Stir Welding.* TWI, Park City, UT, USA, 2003, Paper on CD.

J.K. Doley, S.D. Kore. A study on friction stir welding of dissimilar thin sheets of aluminum alloys AA5052–AA6061. *Journal of Manufacturing Science and Engineering.* 138:114502–114506, 2016.

K. Elangovan, V. Balasubramanian. Influences of pin profile and rotational speed of the tool on the formation of friction stir processing zone in AA2219 aluminum alloy. *Material Science and Engineering A.* 459(1/2):7–18, 2007.

K. Elangovan, V. Balasubramanian. Influences of tool pin profile and tool shoulder diameter on the formation of friction stir processing zone in AA6061 aluminium alloy. *Materials and Design.* 29(2):362–373, 2008.

K. Elangovan, V. Balasubramanian, M. Valliappan. Effect of tool pin profile and tool rotational speed on mechanical properties of friction stir welded AA6061 aluminium alloy. *Materials and Manufacturing Processes.* 23(3):251–260, 2008.

ESAB. Friction Stir Welding—The ESAB Way, ESAB, Korea, 2012.

R. Fu, H. Xu, G. Luan, C. Dong, F. Zhang, G. Li. Top surface microstructure of friction-stir welded AA2524-T3 aluminum alloy joints. *Materials Characterization.* 65:48–54, 2012.

C.B. Fuller, M.W. Mahoney, M. Calabrese, L. Micona. Evolution of microstructure and mechanical properties in naturally aged 7050 and 7075 Al friction stir welds. *Materials Science and Engineering A.* 527:2233–2240, 2010.

A.M. Gaafer, T.S. Mahmoud, E.H. Mansour. Microstructural and mechanical characteristics of AA7020-O Al plates joined by friction stir welding. *Materials Science and Engineering A.* 527:7424–7429, 2010.

W. Gan, K. Okamoto, S. Hirano, K. Chung, C. Kim, R.H. Wagoner. Properties of friction-stir welded aluminum alloys 6111 and 5083. *Journal of Engineering Materials and Technology.* 130:031007-1-031007-15, 2008.

C. Genevois, A. Deschamps, A. Denquin, B.D. Cottignies. Quantitative investigation of precipitation and mechanical behavior for AA2024 friction stir welds. *Acta Materialia.* 53:2447, 2005.

H. Gerard, J.C. Ehrstrom. Friction stir welding of dissimilar alloys for aircraft. *Presented at the 5th International Symposium on Friction Stir Welding,* Metz, France, 2004.

A.S. Golezani, R.V. Barenji, A. Heidarzadeh, H. Pouraliakbar. Elucidating of tool rotational speed in friction stir welding of 7020-T6 aluminum alloy. *International Journal of Advanced Manufacturing Technology.* 81:1155–1164, 2015.

A. Goloborodko, T. Ito, X. Yun, Y. Motohashi, G. Itoh. Friction stir welding of a commercial 7075-T6 aluminum alloy: Grain refinement, thermal stability and tensile properties. *Materials Transactions.* 45(8):2503–2508, 2004.

B. Gungor, E. Kaluc, E. Taban, A. Sik. Mechanical, fatigue and microstructural properties of friction stir welded 5083-H111 and 6082-T651 aluminum alloys. *Materials and Design.* 56:84–90, 2014.

J.F. Guo, H.C. Chen, C.N. Sun, G. Bi, Z. Sun, J. Wei. Friction stir welding of dissimilar materials between AA6061 and AA7075 Al alloys effects of process parameters. *Materials and Design.* 56:185–192, 2014.

N. Guo, Y. Fu, Y. Wang, Q. Meng, Y. Zhu. Microstructure and mechanical properties in friction stir welded 5A06 aluminum alloy thick plate. *Materials and Design.* 113:273–283, 2017.

K. Hassan, P. Prangnell, A. Norman, D. Price, S. Williams. Effect of welding parameters on nugget zone microstructure and properties in high strength aluminium alloy friction stir welds. *Science and Technology of Welding and Joining.* 8:257–268, 2003.

B. Heinz, B. Skrotzki. Characterization of a friction-stir-welded aluminum alloy 6013. *Metallurgical and Materials Transactions B.* 33(3):489–498, 2002.

Honda develops new technology to weld together steel and aluminum and achieves world's first application to the frame of a mass-production vehicle, 6 September, 2012. Available from: http://world.honda.com/

L. Hui-jie, Z. Hui-jie. Repair welding process of friction stir welding groove defect. *Transactions of Non ferrous Metals Society of China.* 19:563–567, 2009.

F.J. Humphreys, M. Hatherly. *Recrystallization and Related Annealing Phenomena,* 1st edn. Elsevier Science Ltd., Oxford, UK, 2002.

F.J. Humphreys, M. Hatherly. *Recrystallization Phenomenon and Related Annealing Phenomena.* Elsevier, Amsterdam, Boston, 2004.

M. Ilangovan, S.R. Boopathy, V. Balasubramanian. Microstructure and tensile properties of friction stir welded dissimilar AA6061–AA5086 aluminium alloy joints. *Transactions of Nonferrous Metals Society of China.* 25:1080–1090, 2015.

K.V. Jata, S.L. Semiatin. Continuous dynamic recrystallization during friction stir welding of high strength aluminum alloys. *Scripta Materialia.* 43:743–749, 2000.

M. Kadlec, R. Ruzek, L. Novakova. Mechanical behaviour of AA 7475 friction stir welds with the kissing bond defect. *International Journal of Fatigue*. 74:7–19, 2015.

S.W. Kallee. Industrial applications of friction stir welding, in: D. Lohwasser and Z. Chen (eds), *Friction Stir Welding from Basics to Applications*. Woodhead Publishing Limited, Cambridge, UK, 118–163, 2010.

S.W. Kallee, J. Davenport, E.D. Nicholas. Railway manufacturers implement friction stir welding. *Welding Journal*. 81:47–50, 2002.

S.W. Kallee, E.D. Nicholas, W.M. Thomas. Industrialisation of friction stir welding for aerospace structures. *Structures and Technologies—Challenges for Future Launchers Third European Conference*, Strasbourg France, December 11–14, 2001.

Ş. Kasman. Multi-response optimization using the Taguchi-based grey relational analysis: A case study for dissimilar friction stir butt welding of AA6082-T6/AA5754-H111. *International Journal of Advanced Manufacturing Technology*. 68:795–804, 2013.

T. Kawasaki, K. Masai. Application of friction stir welding to construction of railway vehicles. *Japan Society of Mechanical Engineers International Journal A*. 47(3):502–511, 2004.

S.H. Kazi, L.E. Murr. Complex flow phenomena associated with friction stir welding of aluminium alloys, in: R.S. Mishra et al. (eds), *Friction Stir Welding and Processing*. TMS, Warrendale, PA, USA, 139–151, 2001.

N.Z. Khan, Z.A. Khan, A.N. Siddiquee. Effect of shoulder diameter to pin diameter (D/d) ratio on tensile strength of friction stir welded 6063 aluminium alloy. *Materials Today: Proceedings*. 2:1450–1457, 2015a.

N.Z. Khan, A.N. Siddiquee, Z.A. Khan, A.K. Mukhopadhyay. Mechanical and microstructural behavior of friction stir welded similar and dissimilar sheets of AA2219 and AA7475 aluminium alloys. *Journal of Alloys and Compounds*. 695:2902–2908, 2017.

N.Z. Khan, A.N. Siddiquee, Z.A. Khan, S.K. Shihab. Investigations on tunneling and kissing bond defects in FSW joints for dissimilar aluminum alloys. *Journal of Alloys and Compounds*. 648:360–367, 2015b.

S.A. Khodir, T. Shibayanagi. Microstructure and mechanical properties of friction stir welded dissimilar aluminum joints of AA2024-T3 and AA7075-T6. *Materials Transactions*. 48(7):1928–1937, 2007.

S.A. Khodir, T. Shibayanagi. Friction stir welding of dissimilar AA2024 and AA7075 aluminum alloys. *Material Science and Engineering B*. 148:82–87, 2008.

W.K. Kim, B.C. Goo, S.T. Won. Optimal design of friction stir welding process to improve tensile force of the joint of A6005 extrusion. *Materials and Manufacturing Processes*. 25:637–643, 2010.

Y.G. Kim, H. Fujii, T. Tsumur, T. Komazaki, K. Nakata. Three defect types in friction stir welding of aluminum die casting alloy. *Material Science and Engineering A*. 415:250254, 2006.

H. Klopstock, A.R. Neelands. An improvement method of joining or welding metals. British Patent specification 572789, 1941.

D. Klotzer, C. Ullner, E. Tyulyukovskiy, N. Huber. Identification of viscoplastic material parameters from spherical indentation data: Part II. Experimental validation of the method. *Journal of Materials Research.* 21(3):677–684, 2006.

M. Koilraj, V. Sundareswaran, S. Vijayan, S.R.K. Rao. Friction stir welding of dissimilar aluminium alloys AA2219 to AA5083: Optimization of process parameters using Taguchi technique. *Materials and Design.* 42:1–7, 2012.

K.N. Krishnan. On the formation of onion rings in friction stir welds. *Material Science and Engineering A.* 327:246–251, 2002.

K. Kumar, S.V. Kailas. On the role of axial load and the effect of interface position on the tensile strength of a friction stir welded aluminum alloy. *Materials and Design.* 29:791–797, 2008a.

K. Kumar, S.V. Kailas. The role of friction stir welding tool on material flow and weld formation. *Material Science and Engineering A.* 485(1):367–374, 2008b.

N.T. Kumbhar, K. Bhanumurthy. Friction stir welding of Al 5052 with Al 6061 alloys. *Journal of Metallurgy.* 1–7, 2012.

R.M. Leal, C. Leitao, A. Loureiro, D.M. Rodrigues, P. Vilaca. Material flow in heterogeneous friction stir welding of thin aluminium sheets: Effect of shoulder geometry. *Materials Science and Engineering A.* 498(1):384–391, 2008.

R.M. Leal, A. Loureiro. Defects formation in friction stir welding of aluminium alloys. *Materials Science Forum.* 455(6):299–302, 2004.

W.B. Lee, Y.M. Yeon, S.B. Jung. Evaluation of the microstructure and mechanical properties of friction stir welded 6005 aluminum alloy. *Materials Science and Technology.* 19(11):1513–1518, 2003a.

W.B. Lee, Y.M. Yeon, S.B. Jung. The joint properties of dissimilar formed Al alloys by friction stir welding according to the fixed location of materials. *Scripta Materialia.* 49(5):423–428, 2003b.

W.B. Lee, S.B. Jung. Void free friction stir weld zone of the dissimilar 6061 aluminium and copper joint by shifting the tool insertion location. *Materials Research Innovations.* 8(2):93–96, 2003c.

C. Leitao, R.M. Leal, D.M. Rodrigues, A. Loureiro, P. Vilaca. Mechanical behaviour of similar and dissimilar AA5182-H111 and AA6016-T4 thin friction stir welds. *Materials and Design.* 30:101–108, 2009.

A.J. Leonard. Microstructure and ageing behaviour of FSWs in aluminium alloys 2014AT651 and 7075-T651. *2nd Friction Stir Welding Symposium*, TWI, Gothenburg, Sweden, 2000.

B. Li, Z. Zhang, Y. Shen, W. Hu, L. Luo. Dissimilar friction stir welding of Ti–6Al–4V alloy and aluminum alloy employing a modified butt joint configuration: Influences of process variables on the weld interfaces and tensile properties. *Materials and Design.* 53:838–848, 2014.

J.Q. Li, H.J. Liu. Effects of welding speed on microstructures and mechanical properties of AA2219-T6 welded by the reverse dual-rotation friction stir welding. *International Journal of Advanced Manufacturing Technology.* 68:2071–2083, 2013.

Y. Li, L.E. Murr, J.C. McClure. Flow visualization and residual microstructures associated with the friction-stir welding of 2024 aluminum to 6061 aluminum. *Materials Science and Engineering A.* 271(1–2):213–223, 1999a.

Y. Li, L.E. Murr, J.C. McClure. Solid-state flow visualization in the friction-stir welding of 2024 Al to 6061 Al. *Scripta Materialia*. 40(9):1041–1046, 1999b.

X. Liang, H. Li, Z. Li, T. Hong, B. Ma, S. Liu, Y. Liu. Study on the microstructure in a friction stir welded 2519-T87 Al alloy. *Materials and Design*. 35:603–608, 2012.

V.M. Linton, M.I. Ripley. Influence of time on residual stresses in friction stir welds in age hardenable 7xxx aluminium alloys. *Acta Materialia*. 56:4319, 2008.

H.J. Liu, H. Fujii, M. Maeda, K. Nogi. Tensile properties and fracture locations of friction-stir welded joints of 6061-T6 aluminum alloy. *Journal of Materials Science Letters*. 22(15):1061–1063, 2003.

H.J. Liu, H. Fujii, K. Nogi. Friction stir welding characteristics of 2017-T351 aluminum alloy sheet. *Journal of Materials Science*. 40:3297–3299, 2005.

X. Liu, S. Lan, J. Ni. Analysis of process parameters effects on friction stir welding of dissimilar aluminum alloy to advanced high strength steel. *Materials and Design*. 59:50–62, 2014.

X.C. Liu, C.S. Wu. Elimination of tunnel defect in ultrasonic vibration enhanced friction stir welding. *Materials and Design*. 90:350–358, 2016.

T. Long, W. Tang, A.P. Reynolds. Process response parameter relationships in aluminium alloy friction stir welds. *Science and Technology of Welding and Joining*. 12(4):311–317, 2007.

O. Lorrain, V. Favier, H. Zahrouni, D. Lawrjaniec. Understanding the material flow path of friction stir welding process using unthreaded tools. *Journal of Materials Processing Technology*. 210:603–609, 2010.

T. Luijendijk. Welding of dissimilar aluminium alloys. *Journal of Materials Processing Technology*. 103:29–35, 2000.

Mazda Develops World's First Steel and Aluminum Joining Technology Using Friction Heat, June 2, 2005. http://www2.mazda.com

T.R. McNelley, S. Swaminathan, J.Q. Su. Recrystallization mechanisms during friction stir welding/processing of aluminium alloys. *Scripta Materialia*. 58:349–354, 2008.

H.J. McQueen, S. Spigareli, M.E. Kassner, E. Evangelista. *Hot Deformation and Processing of Aluminum Alloys*. CRC Press, Boca Raton, FL, USA, 2011.

R.W. Messler. *Principles of Welding: Processes, Physics, Chemistry and Metallurgy*. John Wiley & Sons Inc., New York, 1999.

W.S. Miller, L. Zhuang, J. Bottema, A.J. Witterbrood, P. De Smet, A. Haszler, A. Vieregge. Recent development in aluminium alloys for the automotive industry. *Materials Science and Engineering A*. 280:37–49, 2000.

R.S. Mishra, Z.Y. Ma. Friction stir welding and processing. *Materials Science and Engineering R*. 50:(1–2)1–78, 2005.

P.M.G.P. Moreira, M.A.V. de Figueiredo, P.M.S.T. de Castro. Fatigue behaviour of FSW and MIG weldments for two aluminium alloys. *Theoretical and Applied Fracture Mechanics*. 48(2):169–177, 2007.

L.E. Murr, R.D. Flores, O.V. Flores, J.C. McClure, G. Liu, D. Brown. Friction stir welding: microstructural characterization. *Materials Research Innovations*. 1(4):211–223, 1998.

S. Muthukumaran, S.K. Mukherjee. Multi-layered metal flow and formation of onion rings in friction stir welds. *International Journal of Advanced Manufacturing Technology.* 38:68–73, 2008.

K. Nakata, Y.G. Kim, M. Ushio, T. Hashimoto, S. Jyogan. Weldability of high strength aluminium alloys by friction stir welding. *ISIJ International.* 40:515–519, 2000.

E.D. Nicholas, W.M. Thomas. A review of friction processes for aerospace applications. *International Journal of Materials and Product Technology.* 13(1):45–55, 1998.

C.A.W. Olea, L. Roldo, J.F. dos Santos, T.R. Strohaecker. Friction stir welding of precipitate hardenable aluminium alloys: A review. *IIW—Welding in the World.* 50(11/12):78–87, 2006.

C.A.W. Olea, L. Roldo, J.F. dos Santos, T.R. Strohaecker. A sub-structural analysis of friction stir welded joints in an AA6056 Al-alloy in T4 and T6 temper conditions. *Materials Science and Engineering A.* 454–455:52–62, 2007.

A. Oosterkamp, L.D. Oosterkamp, A. Nordeide. 'KB' phenomena in solid-state welds of aluminum alloys. *Welding Journal.* 83(8):225–231, 2004.

J. Ouyang, E. Yarrapareddy, R. Kovacevic. Microstructural evolution in the friction stir welded 6061 aluminum alloy (T6-temper condition) to copper. *Journal of Materials Processing Technology.* 172(1):110–122, 2006.

G. Padmanaban, V. Balasubramanian. Selection of FSW tool pin profile, shoulder diameter and material for joining AZ31B magnesium alloy: An experimental approach. *Materials and Design.* 30:2647–2656, 2009.

R. Palanivel, P.K. Mathews, N. Murugan, I. Dinaharan. Effect of tool rotational speed and pin profile on microstructure and tensile strength of dissimilar friction stir welded AA5083-H111 and AA6351-T6 aluminum alloys. *Materials and Design.* 40:7–16, 2012.

R. Palanivel, P. Koshy Mathews, N. Murugan. Optimization of process parameters to maximize ultimate tensile strength of friction stir welded dissimilar aluminum alloys using response surface methodology. *Journal of Central South University of Technology.* 20:2929–2938, 2013.

R. Palanivel, P.K. Mathews, I. Dinaharan, N. Murugan. Mechanical and metallurgical properties of dissimilar friction stir welded AA5083-H111 and AA6351-T6 aluminum alloys. *Transactions of Nonferrous Metals Society of China.* 24. 58–65, 2014.

S.K. Park, S.T. Hong, J.H. Park, K.Y. Park, Y.J. Kwon, H.J. Son. Effect of material locations on properties of friction stir welding joints of dissimilar aluminium alloys. *Science and Technology of Welding and Joining.* 15(4):331–336, 2010.

M. Peel, A. Steuwer, M. Preuss, P.J. Withers. Microstructure, mechanical properties and residual stresses as a function of welding speed in aluminium AA5083 friction stir welds. *Acta Materialia.* 51:4791–4801, 2003.

R. Priya, V.S. Sarma, K.P. Rao. Effect of postweld heat treatment on the microstructure and tensile properties of dissimilar friction stir welded AA 2219 and AA 6061 alloys. *Transactions Indian Institute Metals.* 62:11–19, 2009.

I. Radisavljevic, A. Zivkovic, N. Radovic, V. Grabulov. Influence of FSW parameters on formation quality and mechanical properties of Al 2024-T351 butt welded joints. *Transactions of Nonferrous Metals Society of China*. 23:3525–3539, 2013.

S. Rajakumar, C. Muralidharan, V. Balasubramanian. Influence of friction stir welding process and tool parameters on strength properties of AA7075-T6 aluminium alloy joints. *Materials and Design*. 32(2):535–549, 2011.

D. Rao, K. Huber, J. Heerens, J.F. dosSantos, N. Huber. Asymmetric mechanical properties and tensile behaviour prediction of aluminium alloy 5083 friction stir welding joints. *Materials Science and Engineering A*. 565:44–50, 2013.

T.S. Rao, G.M. Reddy, S.R.K. Rao. Microstructure and mechanical properties of friction stir welded AA7075−T651 aluminum alloy thick plates. *Transactions of Nonferrous Metals Society of China*. 25:1770–1778, 2015.

K.H. Rendigs. Aluminium structures used in aerospace-status and prospects. *Materials Science Forum*. 242:11–24, 1997.

C.G. Rhodes, M.W. Mahoney, W.H. Bingel, R.A. Spurling, C.C. Bampton. Effects of friction stir welding on microstructure of 7075 aluminum. *Scripta Materialia*. 36:69–75, 1997.

D.M. Rodrigues, A. Loureiro, C. Leitao, R.M. Leal, B.M. Chaparro, P. Vilaça. Influence of friction stir welding parameters on the microstructural and mechanical properties of AA 6016-T4 thin welds. *Materials and Design*. 30:1913–1921, 2009.

R.I. Rodriguez, J.B. Jordon, P.G. Allison, T. Rushing, L. Garcia. Microstructure and mechanical properties of dissimilar friction stir welding of 6061-to-7050 aluminum alloys. *Materials and Design*. 83:60–65, 2015.

C.E.D. Rowe, W. Thomas. *Advances in Tooling Materials for Friction Stir Welding*. TWI and Cedar Metals Ltd. Internet Publication by TWI, Cambridge, January 13, 2005.

Y.S. Sato, H. Kokawa, M. Enomoto, S. Jogan. Microstructural evolution of 6063 aluminum during friction-stir welding. *Metallurgical and Materials Transactions A*. 30(9):2429–2437, 1999a.

Y.S. Sato, H. Kokawa, M. Enomoto, S. Jogan, T. Hashimoto. Precipitation sequence in friction stir weld of 6063 aluminum during aging. *Metallurgical and Materials Transactions A*. 30(12):3125–3130, 1999b.

Y.S. Sato, Y. Kurihara, H. Kokawa. Microstructural characteristics of dissimilar butt friction stir welds of AA7075 and AA2024. *6th International Symposium on FSW*, Toronto, Canada, October 10–13, 2006.

Y.S. Sato, S.H.C. Park, M. Michiuchi, H. Kokawa. Constitutional liquation during dissimilar friction stir welding of Al and Mg alloys. *Scripta Materialia*. 50:1233–1236, 2004.

W.F. Savage, E.F. Nippes, E.S. Szekeres. Study of weld interface phenomena in a low alloy steel. *Welding Journal*. 55(9):260, 1976.

A. Scialpi, L.A.C. De Filippis, P. Cavaliere. Influence of shoulder geometry on microstructure and mechanical properties of friction stir welded 6082 aluminium alloy. *Materials and Design*. 28:1124–1129, 2007.

A. Scialpi, M. De Giorgi, L.A.C. De Filippis, R. Nobile, F.W. Panella. Mechanical analysis of ultra-thin friction stir welding joined sheets with dissimilar and similar materials. *Materials and Design*. 29:928–936, 2008.

C. Sharma, D.K. Dwivedi, P. Kumar. Effect of welding parameters on microstructure and mechanical properties of friction stir welded joints of AA7039 aluminum alloy. *Materials and Design*. 36:379–390, 2012.

I. Shigematsu, Y. Kwon, K. Suzuki, T. Imai, N. Saito. Joining of 5083 and 6061 aluminum alloys by friction stir welding. *Journal of Materials Science Letters*. 22(5):353–356, 2003.

A.A.M.d. Silva, E. Arruti, G. Janeiro, E. Aldanondo, P. Alvarez, A. Echeverria. Material flow and mechanical behaviour of dissimilar AA2024-T3 and AA7075-T6 aluminium alloys friction stir welds. *Materials and Design*. 32:2021–2027, 2011.

A.C.F. Silva, D.F.O. Braga, M.A.V. de Figueiredo, P.M.G.P. Moreira. Friction stir welded T-joints optimization. *Materials and Design*. 55:120–127, 2014.

A. Simar, Y. Bréchet, B. de Meester, A. Denquin, T. Pardoen. Microstructure, local and global mechanical properties of friction stir welds in aluminium alloy 6005A-T6. *Materials Science and Engineering A*. 486(1–2):85–95, 2008.

M. Sivashanmugam, S. Ravikumar, T. Kumar, V.S. Rao, D. Muruganandam. A review on friction stir welding for aluminium alloys. 978-1-4244-9082-0/10/$26.00 ©2010 IEEE, pp. 216–221.

C.B. Smith, W. Crusan, J.R. Hootman, J.F. Hinrichs, R.J. Heideman, J.S. Noruk. Friction stir welding in automotive industry. In *Proceedings of the TMS-Aluminium Automotive and Joining Sessions*. 2001, pp. 175–185.

S.K. Son, M. Takeda, M. Mitome, Y. Bando, T. Endo. Precipitation behavior of an Al–Cu alloy during isothermal aging at low temperatures. *Materials Letters*. 59:629–632, 2005.

Y. Song, X. Yang, L. Cui, X. Hou, Z. Shen, Y. Xu. Defect features and mechanical properties of friction stir lap welded dissimilar AA2024–AA7075 aluminum alloy sheets. *Materials and Design*. 5:9–18, 2014.

Space Shuttle Technology Summary, *Friction Stir Welding*. NASA, Maryland, 2001

P.B. Srinivasan, W. Dietzel, R. Zettler, J.F. dos Santos, V. Sivan. Stress corrosion cracking susceptibility of friction stir welded AA7075–AA6056 dissimilar joint. *Materials Science and Engineering A*. 392:292–300, 2005.

P.B. Srinivasan, W. Dietzel, R. Zettler, J.F. Santos, V. Sivan. Effects of inhibitors on corrosion behaviour of dissimilar aluminium alloy friction stir weldment. *Corrosion Engineering, Science and Technology*. 42:161–167, 2007.

E.A. Starke, J.T. Staley. Application of modern aluminum alloys to aircraft. *Progress in Aerospace Sciences*. 32:131–172, 1996.

A. Steuwer, M.J. Peel, P.J. Withers. Dissimilar friction stir welds in AA5083–AA6082: The effect of process parameters on residual stress. *Materials Science and Engineering A*. 441:187–196, 2006.

J.Q. Su, T.W. Nelson, R. Mishra, M. Mahoney. Microstructural investigation of friction stir welded 7050-T651 aluminium. *Acta Materialia*. 51:713–729, 2003.

L.E. Svensson, L. Karlsson. Microstructure, hardness and fracture in friction stir welded AA 6082, in: *Proc 1st Int Friction Stir Welding Symp*, Oaks, CA, USA, June 14–16, 1999, Paper on CD.

W. Tang, X. Guo, J.C. McClure, L.E. Murr, A. Nunes. Heat input and temperature distribution in friction stir welding. *Journal of Materials Processing and Manufacturing Science*. 7:163–172, 1998.

S.M.O. Tavares, J.F. dos Santos, P.M.S.T. de Castro. Friction stir welded joints of Al–Li alloys for aeronautical applications: Butt-joints and tailor welded blanks. *Theoretical and Applied Fracture Mechanics*. 65:8–13, 2013.

W.M. Thomas, S.W. Kallee, D.G. Staines, P.J. Oakley. Friction stir welding—Process variants and developments in the automotive industry, *Presented at SAE World Congress*, Cobo Center, Detroit, MI, USA, April 3–7, 2006.

W.M. Thomas, E.D. Nicholas, J.C. Needham, M.G. Murch, P. Templesmith, C.J. Dawes. Patents Friction stir butt welding. G B Patent Application No. 9125978.8, USA, 1991.

P. Threadgill. Friction stir welds in aluminium alloys: Preliminary microstructural assessment. *TWI Bulletin*. 38:30–33, 1997.

D. Venkateswarlu, P.N. Rao, M.M. Mahapatra, S.P. Harsha, N.R. Mandal. Processing and optimization of dissimilar friction stir welding of AA 2219 and AA 7039 alloys. *Journal of Materials Engineering and Performance*. 24(12):4809–4824, 2015.

H. Wang, P.A. Colegrove, J.F. dos Santos. Numerical investigation of the tool contact condition during friction stir welding of aerospace aluminium alloy. *Computational Materials Science*. 71:101–108, 2013.

A.A. Zadpoor, J. Sinke, R. Benedictus. Global and local mechanical properties and microstructure of friction stir welds with dissimilar materials and/or thicknesses. *Metallurgical and Materials Transaction A*. 41:3365–3378, 2010.

R. Zettler, T. Vugrin, M. Schmucker. Effects and defects of friction stir welds, in: D. Lohwasser and Z. Chen (eds), *Friction Stir Welding from Basics to Applications*. Woodhead Publishing Limited, Cambridge, 245–265, 2010.

X.X. Zhang, B.L. Xiao, Z.Y. Ma. A transient thermal model for friction stir weld: Part II. Effects of weld conditions. *Metallurgical and Materials Transaction A*. 42:3229–3239, 2011.

Y.N. Zhang, X. Cao, S. Larose, P. Wanjara. Review of tools for friction stir welding and processing. *Canadian Metallurgical Quarterly*. 51(3):250–261, 2012.

Z. Zhang, B.L. Xiao, Z.Y. Ma. Enhancing mechanical properties of friction stir welded 2219Al-T6 joints at high welding speed through water cooling and post-welding artificial ageing. *Materials Characterization*. 106:255–265, 2015.

Y. Zhao, L. Zhou, Q. Wang, K. Yan, J. Zou. Defects and tensile properties of 6013 aluminum alloy T-joints by friction stir welding. *Materials and Design*. 57:146–155, 2014.

Index

Milton Keynes UK
Ingram Content Group UK Ltd.
UKHW040053071024
449327UK00019B/540